S+SPATIALSTATS

W9-BYG-662

Springer
New York
Berlin
Heidelberg
Barcelona
Hong Kong
London
Milan
Paris
Singapore
Tokyo

Stephen P. Kaluzny
Silvia C. Vega
Tamre P. Cardoso
Alice A. Shelly

S+SPATIALSTATS

User's Manual for Windows® and UNIX®

Springer

Stephen P. Kaluzny, Silvia C. Vega,
     Tamre P. Cardoso, and Alice A. Shelly
MathSoft, Inc.
Data Analysis Products Division
1700 Westlake Avenue North
Suite 500
Seattle, WA 98109-9891 USA

Library of Congress Cataloging-in-Publication Data
S+SpatialStats : user's manual for Windows and UNIX / Stephen P.
   Kaluzny... [et al.].
        p.   cm.
   Includes bibliographical references and index.
   ISBN 0-387-98226-4 (softcover : alk. paper)
   1. S+SpatialStats.   2. Spatial analysis (Statistics)—Data
processing.   3. Microsoft Windows (Computer file)   4. UNIX (Computer
file)   I. Kaluzny, S. P.
QA278.2.S18   1997
519.5'35—dc21   97-10093

Printed on acid-free paper.

© 1998 MathSoft, Inc.
All rights reserved. This work may not be translated or copied in whole or in part without the written permission
of the publisher (Springer-Verlag New York, Inc., 175 Fifth Avenue, New York, NY 10010, USA), except for
brief excerpts in connection with reviews or scholarly analysis. Use in connection with any form of information
storage and retrieval, electronic adaptation, computer software, or by similar or dissimilar methodology now
known or hereafter developed is forbidden.
The use of general descriptive names, trade names, trademarks, etc., in this publication, even if the former are
not especially identified, is not to be taken as a sign that such names, as understood by the Trade Marks and
Merchandise Marks Act, may accordingly be used freely by anyone.

The license management portion of this product is based on Élan License Manager. © 1989-1996 Élan Computer Group, Inc. All Rights Reserved.

S-PLUS and StatSci are registered trademarks and S+GISLINK and S+SPATIALSTATS are trademarks of Math-
Soft, Inc. Trellis, S, and New S are trademarks of Lucent Technologies. UNIX is a registered trademark of
X/Open Company, Ltd. SPARC and Sun are registered trademarks of Sun Microsystems, Inc. Microsoft is a
registered trademark and Windows is a trademark of Microsoft Corporation. Élan License Manager is a trademark of Élan Computer Group.

Production managed by Steven Pisano; manufacturing supervised by Jeffrey Taub.
Photocomposed pages prepared from the authors' LaTeX and nroff files using Springer's svwide macro..
Printed and bound by Maple-Vail Book Manufacturing Group, York, PA.
Printed in the United States of America.

9 8 7 6 5 4 3 2

ISBN 0-387-98226-4 Springer-Verlag New York Berlin Heidelberg SPIN 10738809

Contents

How to Use This Book

This manual describes how to use the S-PLUS spatial statistics module, S+SPATIALSTATS, and includes detailed descriptions of the principal S+SPATIALSTATS functions.

In this book, you will learn how to perform the following tasks with the S+SPATIALSTATS module:

- Install S+SPATIALSTATS on either a UNIX or Windows system.

- Use various S-PLUS and S+SPATIALSTATS functions for exploratory data analysis of spatial data.

- Reduce dimensionality of spatial data using hexagonal binning.

- Use variogram estimation and kriging with geostatistical data.

- Calculate various measures of spatial correlation for lattice data.

- Model lattice data with autoregressive and moving-average models.

- Test spatial point patterns for complete spatial randomness.

- Estimate the intensity and the K-function for regular spatial point patterns.

- Simulate spatial data.

Intended Audience

This book, like the S+SPATIALSTATS module, is intended for geographers, scientists, and other data analysts who need to analyze spatial data. This book assumes a working knowledge of S-PLUS, such as can be obtained from reading *A Gentle Introduction to* S-PLUS or *A Crash Course in* S-PLUS. It is also assumed that the user has a basic knowledge of statistics and in particular, is familiar with spatial statistics. This book is not meant to be a text book in spatial statistics. See the section on related books later in this preface for some recommended references on spatial statistics.

User's Manual Organization

The main body of this book is divided into seven chapters. Chapter 1 provides some introductory material on spatial data and S+SPATIALSTATS, including some definitions and an overview of the basic spatial analysis tools provided with S+SPATIALSTATS. Chapter 2 is also introductory, but focuses on starting and quitting S+SPATIALSTATS, setting up working directories in S-PLUS, and importing/exporting spatial data. The remaining chapters cover spatial data analysis for three broad categories of spatial data: geostatistical data, lattice data, and spatial point patterns (see sections 1.1 and 1.3 for definitions). Chapter 3 introduces and illustrates the use of S-PLUS and S+SPATIALSTATS functions for exploratory data analysis and visualization of spatial data. Specific examples for the three types of spatial data are included. Chapters 4 through 6 introduce the use of S+SPATIALSTATS functions for the analysis of geostatistical data, lattice data, and spatial point patterns, respectively. Chapter 7 is intended for ARC/INFO users, and defines the correspondence between ARC/INFO coverage data and appropriate spatial analysis methods. Specific examples of spatial data analysis for several coverage types are included.

This book also includes four appendices: appendix A has installation intructions for S+SPATIALSTATS; appendix B provides a list of the S+SPATIALSTATS functions and data sets organized by categories; appendix C contains help files documenting the sample data sets provided with S+SPATIALSTATS; and appendix D contains individual help files for S+SPATIALSTATS functions. The appendices are followed by a glossary of commonly used terms in spatial statistics.

Typographic Conventions

This book uses the following typographic conventions:

- The *italic font* is used for emphasis, and also for user-supplied variables within UNIX, DOS, and S-PLUS commands.

- The **bold font** is used for UNIX and DOS commands and filenames, as well as for chapter and section headings. For example,

  setenv S_PRINT_ORIENTATION portrait
  SET SHOME=C:\SPLUS

  In this font, both " and " represent the double-quote key on your keyboard (").

- The `typewriter font` is used for S-PLUS functions and examples of S-PLUS sessions. For example,

  ```
  > plot(bramble)
  ```

  Displayed S-PLUS commands are shown with the S-PLUS prompt `>`. Commands that require more than one line of input are displayed with the S-PLUS continuation prompts `+` and `Continue string:`.

- Boxed text represents either keys from the workstation keyboard or mouse buttons. For example,

  To delete a character, press the Delete key.

 Warning: The "dangerous bend" sign followed by the word **Warning** indicates a warning about S-PLUS behavior. Read these warnings carefully.

⟹   **Hint:** When you see the right arrow followed by the word **Hint**, you are getting a peek ahead into more sophisticated use of S-PLUS.

Note: Points of interest that are neither warnings nor hints are preceded by the word **Note**.

Related Books

For users familiar with S-PLUS, the S+SPATIALSTATS *User's Manual* contains all the information most users need to begin making productive use of S+SPATIALSTATS. Users who are *not* familiar with S-PLUS should read one of the following manuals to gain a working knowledge of S-PLUS:

- *A Gentle Introduction to* S-PLUS is a step-by-step guide to the most basic S-PLUS operations. It is intended to help new users who are unfamiliar with both the operating system and data analysis software get started using S-PLUS.

- *A Crash Course in* S-PLUS is a fast-paced tour of the capabilities of S-PLUS. It is intended for users who are familiar with other data analysis packages and want to quickly master S-PLUS syntax and work with specific S-PLUS functions.

- The S-PLUS *User's Manual* provides complete procedures for the basic operation of S-PLUS, including graphics manipulation, customization, and data input and output.

Other useful information can be found in the The S-PLUS *Guide to Statistical and Mathematical Analysis*. This manual describes how to analyze data using a variety of statistical and mathematical techniques, including classical statistical inference, time series analysis, linear regression, ANOVA models, generalized linear and generalized additive models, loess models, nonlinear regression, and regression and classification trees.

For general, broad references in the field of spatial statistics, we refer the user to:

- Cressie, Noel A. C. (1993). *Statistics for Spatial Data*, Revised Edition. John Wiley and Sons, New York.

- Ripley, B. D. (1981). *Spatial Statistics*. Wiley, New York.

For references specific to given topics in spatial statistics, see below.

For geostatistical data:

- Isaaks, E. H. and Srivastava, R. M. (1989). *An Introduction to Applied Geostatistics*. Oxford University Press, New York.

- Journel, A. G. and Huijbregts, C. J. (1978). *Mining Geostatistics*. Academic Press, London.

For lattice data:

- Cliff, A. D., and J. K. Ord (1981). *Spatial Processes: Models and Applications*. Pion Limited, London.

- Haining, Robert (1993). *Spatial Data Analysis in the Social and Environmental Sciences*. Cambridge University Press, Cambridge.

For spatial point patterns:

- Diggle, Peter J. (1983). *Statistical Analysis of Spatial Point Patterns.* Academic Press Inc., New York.

Many other references on topics in spatial statistics are found in other books and papers included in the bibliography at the end of this manual.

S-PLUS Technical Support

If you have purchased S+SPATIALSTATS in the last 60 days, or if you have purchased a support contract for S+SPATIALSTATS and you have any problems installing or using the product, you can contact S-PLUS technical support in any of the following ways:

For North America:

- **By Electronic Mail:** Send your questions to the following address:

  support@statsci.com

- **By Fax:** Fax your questions to **(206) 283-6310**.

- **By Phone:** Call us at **(206) 283-8802 x235** Monday through Friday from 7:30 a.m. to 5:00 p.m. Pacific Time.

For international customers: Please contact your local distributor.

Comments?

We welcome your comments on this book. Please send electronic mail to the following addresses:

spk@statsci.com (Stephen Kaluzny)
silvia@statsci.com (Silvia Vega)

Acknowledgements

The Research and Development was partially supported by the NASA Earth Observation Commercial Applications Program (EOCAP 93) Contract No. NAS13-623. The software and help file documentation for the spatial linear models, spatial correlation, nearest neighbors, and associated S-PLUS functions were originally developed by Douglas B. Clarkson (Principal Investigator) and Hubert Jin (Investigator) as part of National Institutes of Health Small Business Innovative Research (NIH SBIR) Phase I Grant 1R43CA65340-01.

Contributions to the manual and software were made by others in the MathSoft — Seattle Development Group:

- Bill Dunlap provided valuable design assistance and was always available to help with those difficult programming problems.

- Richard Calaway provided technical editing for the original version of this manual. Jagrata Minardi formatted and helped edit the current version of this book.

The following consultants were major contributors to the design and development of the S+SPATIALSTATS:

- Brian Ripley (Oxford University) provided valuable input on the design and thorough reviews of early versions of the software and its documentation.

- Dan Carr (George Mason University) provided the hexagonal binning functions.

Miles Logsdon with the EOS-Amazon Modeling Project, a NASA supported inter-disciplinary science team at the University of Washington, provided the monthly rainfall data. Peter Guttorp (University of Washington) provided the earthquake data sets. The beta testers for S+SPATIALSTATS provided many helpful comments on the module.

1

Introduction to Spatial Data and S+SPATIALSTATS

S+SPATIALSTATS is an add-on module to the S-PLUS system for data analysis and graphics. S+SPATIALSTATS provides a comprehensive suite of tools designed for the statistical analysis of spatial data. In this chapter you will be introduced to the following:

- Types of spatial data (section 1.1)

- Analysis of spatial data (section 1.2)

- Stochastic model for spatial data (section 1.3)

- Tools available in S+SPATIALSTATS (section 1.4)

- Limitations of S+SPATIALSTATS (section 1.5).

1.1 Types of Spatial Data

Spatial data consist of measurements or observations taken at specific locations or within specific regions. In addition to values for various attributes of interest, spatial data sets also include the locations or relative positions of the data values. Locations may be *point* or *areal* referenced. For example, point referenced data are observations recorded at specific fixed locations and might be referenced by latitude and longitude. Areal referenced data are observations specific to a region; for example, the number of burglaries occurring in census tracts, where each census tract is a region. In both cases, spatial locations may be regular or irregular: point locations may fall on a regularly spaced grid, or may be irregular with varying distances between points; areal locations can comprise equally sized contiguous blocks that might occur in an agriculture field study, or may be of variable size and shape such as the city limits within a county. Spatial data may be continuous, such as the measurements of ore content from a core sample, or discrete, such as the number of measles cases reported by county. Further, the locations may come from a spatial continuum such as the point locations within a mining field, or a discrete set, such as the counties within a state.

S+SPATIALSTATS provides tools for analyzing three specific classes of spatial data: geostatistical data, lattice data, and spatial point patterns.

1.1.1 Geostatistical Data

Geostatistical data, also termed random field data, are measurements taken at fixed locations. The locations are generally spatially continuous. Examples of continuous geostatistical data include mineral concentrations measured at test sites within a mine, rainfall recorded at weather stations, concentrations of pollutants at monitoring stations, and soil permeabilities at sampling locations within a watershed. An example of discrete geostatistical data is count data, such as the number of scallops at a series of fixed sampling sites along the coast.

Geostatistical data often exhibit small-scale variation that may be modeled as spatial correlation. The spatial variability is modeled as a function of the distance between sampling sites, where the sites closer together in space generally have more similar data values. See chapters 3 and 4 for descriptions of methods and tools used in the exploration and analysis of geostatistical data.

1.1.2 Lattice Data

Lattice data are observations associated with spatial regions, where the regions can be regularly or irregularly spaced. The spatial regions can be any spatial collection, and are not limited to a grid. Generally, neighborhood information for the spatial regions is available. An example of regular lattice data is information obtained by remote sensing from satellites. The earth's surface is divided into a series of small rectangles (pixels) and the data are received as a regular lattice in $\mathcal{R}^2$. An example of irregular lattice data is cancer rates corresponding to each county in a state.

Mathematically, a lattice is defined by a set of vertices and edges. The sites form the vertices, which are then connected to neighboring sites by edges.

Since lattice data are defined for spatial regions, a method of referencing sites must be determined; sites are often referenced by the centroids of the regions. Figure 1.1 shows an example of a lattice for community medicine areas (CMAs) in Glasgow, Scotland (data taken from Table 7.10, Haining (1990)).

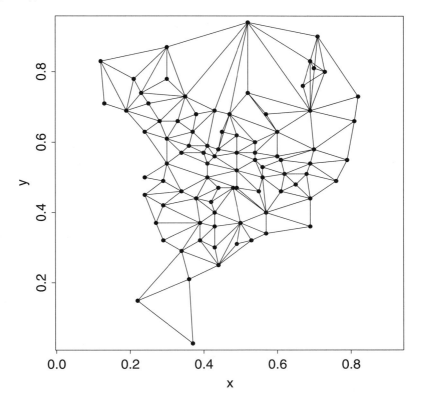

FIGURE 1.1. Glasgow lattice.

A lattice is composed of an index set of sites with an associated set of neighbors. For the Glasgow data, the set of sites consists of CMAs: $1, 2, ..., 87$. The sites are represented by CMA centers which correspond to the vertices in figure 1.1. The set of neighbors can be defined by distance, common boundaries, or other characteristics, depending on the correlation structure of the data. In the representation shown in figure 1.1, the edges denote neighbors defined as adjacent sites. Neighbor relationships can also be weighted; weights based on the length of common boundaries or weights based on the distance from a given site are two examples. See chapters 3 and 5 for descriptions of methods and tools used in the exploration and analysis of lattice data.

1.1.3 Spatial Point Patterns

Point pattern data arise when locations themselves are the variable of interest. Spatial point patterns consist of a finite number of locations observed in a spatial region. Identification of spatial randomness, clustering, or regularity is often the first analysis performed when looking at point patterns. Examples of point pattern data include locations of a species of tree in a forested region, and locations of earthquake epicenters. An example of point pattern data is shown in Figure 1.2.

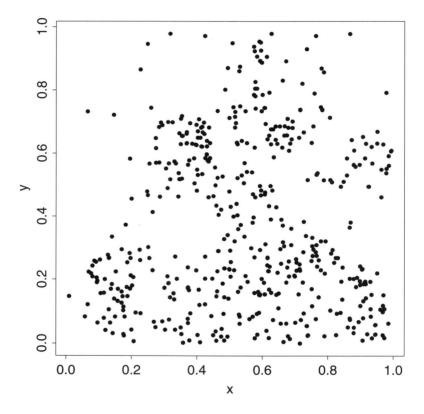

FIGURE 1.2. Locations of Maple trees in a forested region of Lansing Woods, Michigan.

A *marked* spatial point pattern includes values of additional related variables at each location. The additional variables are often called mark variables and may be used to further refine the analysis of point patterns. The Lansing Woods data are a marked spatial point pattern; in addition to the locations, the tree species were also recorded.

1.2   Analysis of Spatial Data

The analysis of spatial data is differentiated from typical data analysis by the inclusion of spatial information in models and predictions. Data are often correlated in space; spatial structure can arise from several different sources, such as measurement error, continuity effects including spatial heterogenity, and space-dependent processes or mechanisms (Haining, 1990). The resulting correlation or covariance structure can be evaluated and used to increase the accuracy of modeling and prediction efforts.

1.2.1   Data Model

Spatial data can often be decomposed into two main components of variation as follows:

data = large-scale variation + small-scale variation.

Large-scale variation may be composed of inter-point or inter-region characteristics representing global trends or gradients. Large-scale variation may also be modeled as local patterns of dependency. The small-scale variation can be thought of as an error term and may be characterized by sources such as measurement error, within region variability, and inherent site variability (Haining, 1990). Models for spatial structure can be incorporated into either component of variation.

1.2.2   Data Collection and Accuracy

As with most collected data, spatial data may have measurement errors and limited accuracy. Unique to spatial data, however, are potential errors associated with the recording of point locations or the representation of areal boundaries. Areal regions, in particular, can be of many sizes and shapes; the method and resolution used to represent the regions in a database may be important as potential sources of error.

Models of spatial dependence are area dependent. They can be affected by factors such as uneven spatial coverage during sampling, as well as the presence of local spatial outliers. Since most spatial dependencies are two-dimensional, consideration of boundary conditions (edge effects) can be more complicated with spatial data. The boundary of a spatial data set extends around an area, affecting a much larger proportion of sites than in a one-dimensional analysis such as time series analysis.

1.2.3 Stationarity

In most cases, a spatial data set represents a single realization of a random process. As such, some degree of *stationarity* must be assumed in order to make inferences about the data. Stationarity refers to some form of *location invariance* of the data. Location invariance implies that the relationships within any subset of points remain the same no matter where the points reside in space. Stationarity may apply to both the mean and the variance. Three common forms of stationarity are:

1. Strict stationarity—requires an equivalence of distribution functions under translation and rotation.

2. Weak stationarity—requires a constant mean and a covariance that is independent of location. The covariance is only dependent on the distance (and perhaps direction) between points. This form requires the existence of a positive, finite variance.

3. Stationarity of increments—requires that the variance of *increments* are independent of location. An increment is defined as the first-order difference between two points.

Stationarity of a process can also be defined in terms of *local* and *global* areas. While a process may be globally nonstationary, the process may exhibit stationarity over local subsets of the global area. Global stationarity of a process may be difficult to verify, while local stationarity may be easier to verify, and sufficient for data analysis.

1.2.4 Isotropy

Since most spatial data is referenced in two dimensions, *isotropy* of any spatial process is important. In this case, isotropy refers to a spatial process that evolves the same in all directions. A spatial process is termed *anisotropic* if the resulting correlation or covariance differs with direction. In many cases, analysis methods for spatial data assume that the spatial correlation is isotropic.

1.2.5 Scale

Scale or spatial resolution is an issue in spatial data analysis. Large-scale patterns observed in spatial data may be the result of different processes operating at different scales. Exploratory analysis of a spatial data set may help to detect patterns at different scales. Patterns discerned from data, however, are dependent on the resolution at which the data were collected.

1.3   Stochastic Model for Spatial Data

Mathematically, many spatial processes can be characterized by a simple stochastic model (Cressie, 1993). Let $\mathbf{s} \in \mathcal{R}^d$ represent a generic data location in d-dimensional Euclidean space. Define the potential datum, $\mathbf{Z}(\mathbf{s})$ at spatial location $\mathbf{s}$, as a random quantity. The multivariate random field (RF),

$$\{\mathbf{Z}(\mathbf{s}) : \mathbf{s} \in D\},$$

is generated by letting $\mathbf{s}$ vary over the index set $D \subset \mathcal{R}^d$.

In practice, a given spatial data set often represents only one realization of an underlying stochastic process. An individual realization of $\mathbf{Z}(\mathbf{s})$ is denoted by $\{z(\mathbf{s}) : \mathbf{s} \in D\}$.

The three distinct types of spatial data previously introduced can be considered as special cases of the general stochastic model as follows:

1. Geostatistical data—D is a fixed subset of $\mathcal{R}^d$ that encompasses a d-dimensional rectangle of positive volume. $\mathbf{Z}(\mathbf{s})$ is a random vector at location $\mathbf{s} \in D$, where the spatial index $\mathbf{s}$ can vary continuously over a subset of D. This is the general characteristic that distinguishes geostatistical data from lattice and point pattern data.

2. Lattice data—D is a fixed collection of countably many spatial sites in $\mathcal{R}^d$. The lattice defined by the index set of sites, D, is supplemented with neighborhood information.

3. Spatial point patterns—D is a point pattern in $\mathcal{R}^d$ or a subset of $\mathcal{R}^d$. Here D is a random set; the index set of D forms a collection of random events that are the spatial point pattern. For the general spatial point pattern, $\mathbf{Z}$ is not specified or is thought of as the scalar $Z(\mathbf{s}) \equiv 1$, for all $\mathbf{s} \in D$. For a marked spatial point pattern, $\mathbf{Z}(\mathbf{s})$ is a random vector at location $\mathbf{s} \in D$.

1.4   Spatial Analysis Tools in S+SPATIALSTATS

S+SPATIALSTATS provides a collection of functions that may be used to perform spatial analyses for various types of spatial data. Additionally, there are many tools available in S-PLUS that are useful for analyzing spatial data, particularly when performing exploratory data analysis. See chapter 3 for specific applications of S-PLUS functions in the exploratory data analysis of spatial data.

For this book, we have separated most of the S+SPATIALSTATS functions into the three categories of spatial data defined in section 1.1. These categories are not meant to be exclusive, since many of the functions can be used on several types of spatial data. For example, spatial regression modeling is typically used for lattice data, but it can be used on geostatistical data. Likewise, it is possible to perform geostatistical-type analyses, such as variogram estimation, on lattice data.

For analyzing geostatistical data, S+SPATIALSTATS includes tools to perform the following:

- Estimate and visualize standard or robust, omnidirectional or directional variograms.

- Model empirical variograms; fit theoretical variogram models to empirical data.

- Perform ordinary kriging to obtain point estimates for unsampled locations and kriging prediction variances.

- Perform universal kriging to model large-scale trend while calculating predictions.

For the analysis of lattice data, S+SPATIALSTATS includes functions to perform the following:

- Find nearest neighbors or groups of neighbors based on distance or common boundaries.

- Calculate and test for spatial autocorrelation using the Moran and Geary correlation coefficients.

- Perform spatial regression modeling using conditional autoregressive, simultaneous autoregressive, or moving-average covariance structures.

For the analysis of point pattern data, S+SPATIALSTATS includes functions to perform the following:

- Plot geometrically accurate point maps without spatial distortion.

- Explore complete spatial randomness using nearest neighbor methods.

- Estimate intensity using kernel, loess, and gaussian methods.

- Test second-order stationarity using Ripley's K-function.

S+SPATIALSTATS also includes methods for simulating the three basic types of spatial data.

1.5   Limitations

S+SPATIALSTATS Version 1.0 was designed to analyze two-dimensional spatial data. Techniques for three-dimensional spatial data are not included in this version, although many of the included techniques can be extended. Also, S+SPATIALSTATS does not include any specific methods for analyzing spatial/temporal data.

S+SPATIALSTATS is not an image processing package. Image processing usually involves specialized data types (e.g. one-byte data) and methods for handling large data sets. S+SPATIALSTATS and the S-PLUS system in general are not designed for this application.

Finally, the S-PLUS system is an interpreted functional programming language. The flexibility it provides comes at the expense of computer memory. A typical analysis requires memory for six or seven copies of your largest data object. This may limit the size of data sets that can be analyzed.

2

Getting Started with S+SPATIALSTATS

If you have used S-PLUS before, getting started with S+SPATIALSTATS is easy; just start S-PLUS, attach the S+SPATIALSTATS module, and begin learning how to use the S+SPATIALSTATS functions—a set of S-PLUS functions for visualizing and analyzing spatial data. This chapter describes the following tasks particular to running the S+SPATIALSTATS module:

- Starting and quitting S+SPATIALSTATS including setting up your environment so that S+SPATIALSTATS is started whenever you start S-PLUS.

- Getting help for S+SPATIALSTATS in the Windows environment. (In the UNIX environment, getting help for S+SPATIALSTATS functions is no different from getting help for other functions in S-PLUS.)

- Using Trellis graphics in S+SPATIALSTATS for the visualization of spatial data.

- Importing your own spatial data into S-PLUS for use with S+SPATIALSTATS.

- Accessing sample data sets included with S+SPATIALSTATS.

If you have not used S-PLUS before, read *A Gentle Introduction to* S-PLUS or *A Crash Course in* S-PLUS to learn the basics of S-PLUS before proceeding with this book. Although both manuals are aimed at new users, *A Gentle Introduction to* S-PLUS is specifically for the user who is new to either computing in general or data analysis in particular, while *A Crash Course in* S-PLUS is specifically for users with experience analyzing data using other software packages. This book assumes only a working knowledge of S-PLUS such as can be gained from reading *A Gentle Introduction to* S-PLUS.

Some of the procedures in this chapter vary depending on whether you run S+SPATIALSTATS under Windows or under UNIX. Where commands or procedures vary between the two operating systems, this chapter describes both.

Note: If you are looking for instructions on installing S+SPATIALSTATS see appendix A.

2.1   Starting and Quitting S+SPATIALSTATS

Start S+SPATIALSTATS by starting S-PLUS, and then attaching the S+SPATIALSTATS functions to your S-PLUS session. To start S-PLUS on a UNIX system, use the command **Splus** from your shell prompt. To start S-PLUS on a Windows system: from Windows 3.1 click on the S-PLUS for Windows icon in Program Manager; from Windows 95 select S-PLUS for Windows from the Start Menu. See *A Gentle Introduction to* S-PLUS or *A Crash Course in* S-PLUS for more detailed instructions on starting S-PLUS.

To add the S+SPATIALSTATS functions to your S-PLUS session, use the following expression:

```
> module(spatial)
```

Note: Adding the S+SPATIALSTATS module causes the `Matrix` library to be attached.

 Warning: The S+SPATIALSTATS module masks some functions in S-PLUS. The masked functions have been modified to support the S+SPATIALSTATS functions, but the changes should not affect normal use of S-PLUS.

To remove the S+SPATIALSTATS module from your S-PLUS session, use the following expression:

```
> module(spatial, unload=T)
```

If you plan to use S+SPATIALSTATS extensively, you may want to customize your S-PLUS startup routine to automatically attach the S+SPATIALSTATS module. You can do this easily by adding the line `module(spatial)` to your `.First` function. If you do not already have a `.First` function, you can create one as follows:

```
> .First <- function(){module(spatial)}
```

To help you keep track of the data that you analyze with S+SPATIAL-STATS, you can create separate directories for individual projects. Each project level directory should have an S-PLUS .Data (or _data on Windows) subdirectory. The .First file defined above should be included in each .Data (or _data on Windows) directory from which you intend to run S+SPATIALSTATS. To work on a particular project, just start S-PLUS from the directory created for that project. How you organize and use your data directories depends on whether you are on a UNIX system or a Windows system. The particular procedures for your system are described in the following sections.

2.1.1 Organizing Your Working Data in UNIX

To create and use the directory *dir* for a S+SPATIALSTATS project in UNIX, use the following commands (% represents the shell prompt; do not type it when issuing the commands):

% **mkdir** *dir dir*/**.Data**
% **cd** *dir*
% **Splus**

In your *first* session in this project directory, use the `module` function directly to attach the S+SPATIALSTATS functions and define the .First function so that the S+SPATIALSTATS module is automatically attached in subsequent sessions:

```
> module(spatial)
> .First <- function(){module(spatial)}
```

2.1.2 Organizing Your Working Data in Windows 3.1

On Windows systems, "the directory where you start S-PLUS" is the directory specified as the Working Directory for the S-PLUS for Windows icon in the Properties dialog in Program Manager. You may want to define a new S-PLUS for Windows icon for each project, each pointing to a different Working Directory.

Note: In S-PLUS, the *working directory* means the directory in which S-PLUS saves your data objects, that is, your **_data** directory. The Working Directory specified in the Properties dialog on Windows 3.1 is referred to in S-PLUS documentation as either the *current directory* or "the directory where you started S-PLUS."

To create a new icon to start S-PLUS for Windows from a different project directory in Windows 3.1:

1. Using the File Manager, create a **_data** subdirectory under the project directory. For example, if you have a directory called **c:\project1** for storing your project1 information, create a new subdirectory named **c:\project1_data**.

2. From Program Manager, highlight your S-PLUS for Windows icon.

3. Choose Copy... from the Edit menu. A dialog box titled Copy Program Item appears, showing the name of the icon to be copied (typically S-PLUS for Windows) and the name of the Program Group to which it belongs (also typically S-PLUS for Windows).

4. Under the heading To Group there is a drop-down list box. Select the same group (say, S-PLUS for Windows) as appears in the From Program Group field.

5. Click OK or press ⌨Enter. A copy of your S-PLUS for Windows icon appears highlighted in the S-PLUS for Windows Program Group.

6. Choose Properties from the File menu.

7. Type a new name in the text field labeled Description; this step allows you to distinguish your new icon from the old.

8. Type the complete path specification of your project directory (*not* of the **_data** directory!) in the text field labeled Working Directory. For example, type **c:\project1**.

9. Click OK or press ⌨Enter.

To use the newly created program icon, simply double-click on it. S-PLUS for Windows starts up using the project directory's **_data** directory as its S-PLUS working directory.

In your *first* session in a new project directory, use the `module` function directly to attach the S+SPATIALSTATS functions and define the `.First` function so that the S+SPATIALSTATS module is automatically attached in subsequent sessions:

```
> module(spatial)
> .First <- function(){module(spatial)}
```

2.1.3   Organizing Your Working Data in Windows 95

In Windows 95, S-PLUS data objects are stored in the **_data** subdirectory of the "Start in" directory for the S-PLUS for Windows icon. You may want to define a new S-PLUS for Windows icon for each project, each pointing to a different Working Directory.

To create a new icon to start S-PLUS for Windows from a different project directory in Windows 95:

1. Create a **_data** subdirectory under the project directory where you want to start S-PLUS. For example, create a new subdirectory named **c:\spluswin\home\project1_data**.

2. From the Start Menu, select Settings for Taskbar. Open the Start Menu Programs screen, and choose Advanced.

3. On the screen that appears, open the S-PLUS for Windows directory under Programs.

4. Create a duplicate copy of the S-PLUS for Windows icon using the Edit menu. Change the name of this icon to the name you want appear on the Startup menu. For example, click the right mouse button and type: **\Project 1**.

5. Highlight the new icon and click the right mouse button, choosing Properties from the drop-down menu.

6. On the page labeled Shortcut, type the complete path specification of your project directory in the text field labeled Start in. (Not in the text field labeled Target!) There must be a subdirectory named **_data** under this directory. For example, type: **c:\spluswin\home\project1**.

7. Click OK or press ⎗Enter⎖.

To use the newly created program icon, simply choose it from the Start menu. S-PLUS for Windows starts up using the project directory's **_data** directory as its S-PLUS working directory.

In your *first* session in a new project directory, use the `module` function directly to attach the S+SPATIALSTATS functions and define the `.First` function so that the S+SPATIALSTATS module is automatically attached in subsequent sessions:

```
> module(spatial)
> .First <- function(){module(spatial)}
```

2.2 Getting Help for S+SPATIALSTATS Functions in Windows

S-PLUS provides online help files for virtually all built-in functions (some functions intended only for internal use have no help files). In S-PLUS, and in S+SPATIALSTATS under UNIX, you obtain help on the function *fun* by using the **help** function as follows:

> `help(`*fun*`)`

However, if you type this expression in S-PLUS for Windows, and *fun* is an S+SPATIALSTATS function, S-PLUS for Windows displays the Search dialog from Windows Help, because the S+SPATIALSTATS functions are not contained in the standard S-PLUS for Windows help file. To obtain help for S+SPATIALSTATS functions in S-PLUS for Windows, you must specify the *module* to which the function belongs when you request help for that function. Thus, to obtain help on the **variogram** function, use **help** as follows:

> `help(variogram, module = "spatial")`

To display the index of S+SPATIALSTATS help files, omit the function name:

> `help(module = "spatial")`

You can select the subject you need help on from the table of contents that appears. The S+SPATIALSTATS help files are also included in appendices C and D of this book.

2.3 Graphics Devices

To use graphics in S-PLUS, you must start one or more *graphics device drivers*. A graphics device can be a window on the screen, a physical printer or plotter, or a software program that controls a physical graphics device. In this manual, all plotting commands assume that a graphics device is open and active. In most cases, we use **motif** or **trellis.device**, depending on the application. If you are using Windows, you need to use **win.graph**, or **trellis.device**. Graphics windows are closed automatically when you exit S-PLUS using **q()**, and you can close a graphics device during a session with **graphics.off** or **dev.off**. For more information on graphics devices in S-PLUS, see the S-PLUS *User's Manual*.

2.4   Using Trellis Graphics in S+SPATIALSTATS

Many of the examples throughout the S+SPATIALSTATS *User's Manual* use Trellis graphics, in conjunction with standard S-PLUS graphics, for displaying spatial data. Additionally, some of the S+SPATIALSTATS functions call Trellis functions directly to display results. Trellis displays are plots which contain one or more panels, arranged in a regular grid-like structure of columns, rows, and pages. Each panel displays a subset of the data, determined by the values of the *given* variables.

Trellis graphics are general enough to encompass a wide variety of 2-D and 3-D displays: histograms, scatter plots, dot plots, contour plots, wireframes, 3-D point clouds and more. However, all panels in a Trellis display must be alike. The data subsets are chosen in a regular manner, conditioning on continuous or discrete variables in the data, thus providing a coordinated series of views of high-dimensional data. The Trellis software gives flexible control over axes and aspect ratios, and contains "banking" computations that let the data select the aspect ratio.

The Trellis library is automatically attached when the S+SPATIALSTATS module is attached, allowing you access to all Trellis functions. Please refer to the S-PLUS *Trellis Displays User's Manual* for specific information on how to use Trellis graphics.

2.5   Importing and Exporting Spatial Data

The examples in this book use spatial data that either are included as S-PLUS data frames or can be generated using the S+SPATIALSTATS functions. For practical use of the S+SPATIALSTATS module, you need to read your own data into S-PLUS. In addition to using the S-PLUS **scan** or **read.table** functions for reading in data stored in ASCII files with a fixed number of fields, S+SPATIALSTATS includes functions to read in some specific types of spatial data. This section describes how to read ASCII files that may contain a variable number of fields with spatial neighborhood information, and how to import/export data from/to GEO-EAS and ARC/INFO.

2.5.1   Reading ASCII Files with Spatial Neighbor Information

The analysis of lattice data involves the use of spatial neighbors defined for each element of a lattice. Generally the number of neighbors per region varies; thus, a file with neighbor identifiers may contain a variable number of fields. The S+SPATIALSTATS function `read.neighbor` can be used to read in ASCII files containing neighbor identifiers. See section 5.1 for details on the use of this function.

2.5.2   Importing and Exporting GEO-EAS Data

GEO-EAS is a DOS based software package that contains a collection of interactive tools for performing two-dimensional geostatistical analyses of spatially distributed data (Englund and Sparks, 1992). GEO-EAS uses a particular format for importing ASCII data. Several other PC software packages for spatial analysis use data in this GEO-EAS file format. Data generated in GEO-EAS can be imported into S-PLUS using the S+SPATIALSTATS `read.geoeas` function as follows:

```
> read.geoeas(file)
```

Here, *file* is a character string containing the name of the GEO-EAS data file to be read. A data frame with as many columns as variables in the file is created. The title line and any measurement units listed with the variable names are ignored.

S+SPATIALSTATS also includes the function `write.geoeas` to write a data frame from S-PLUS to a GEO-EAS formatted ASCII file. The user can optionally specify a file name, title, units for each of the variables, and an end-of-line character. See the help file for more information about this function.

2.5.3   Importing and Exporting ARC/INFO Data

The S-PLUS product S+GISLINK provides tools for transferring data between ARC/INFO and S-PLUS. S+GISLINK provides functions to transfer data stored in coverages and auxiliary info tables. See the *S+GISLINK User's Manual* for more details.

2.6   Sample Data Sets Available in S+SPATIALSTATS

Many of the spatial data sets that are used in examples throughout this manual are included with S+SPATIALSTATS as S-PLUS objects. The data sets are documented in appendix C. These data set help files are also available through on-line help.

3

Visualizing Spatial Data

This chapter introduces exploratory data analysis (EDA) and visualization techniques for spatial data in S-PLUS and S+SPATIALSTATS. EDA involves methods of describing data and its structure in order to formulate hypotheses and check the validity of assumptions. In general, we will be concerned with the distribution of data and violations of local and global stationarity (as defined in chapter 1)—including trends and outliers. Although a researcher uses EDA techniques throughout spatial analysis and modeling, this chapter focuses on initial explorations—before analysis has been attempted. Analysis techniques for specific types of spatial data—geostatistical data, lattice data, and spatial point patterns, will be discussed in chapters 4, 5, and 6.

In this chapter you will learn to do the following tasks:

- Use basic tools of EDA in S-PLUS (section 3.1).

- Apply basic EDA techniques to geostatistical data (section 3.2).

- Apply basic EDA techniques to lattice data (section 3.3).

- Apply basic EDA techniques to point pattern data (section 3.4).

- Apply hexagonal binning (section 3.5).

3.1 EDA Tools

Most of the techniques presented in this section are available in the regular S-PLUS package, so the reader familiar with S-PLUS may want to skip to the examples in subsequent sections. The three main types of EDA techniques introduced in this section are:

- Descriptive Statistics (section 3.1.1)

- Plots and Graphics (section 3.1.2)

- Classification and Clustering Methods (section 3.1.3).

3.1.1 Descriptive Statistics

Descriptive statistics allow a first look at the characteristics of your data. They include simple ways to look for outliers and departures from normality. We can easily obtain summary statistics with the S-PLUS function summary, illustrated for the data frame aquifer:

```
> summary(aquifer)
```

```
        easting              northing              head
Min.    :-145.20    Min.    :  9.414    Min.    :1024
1st Qu.: -21.30    1st Qu.: 33.680    1st Qu.:1548
Median :  11.66    Median : 59.160    Median :1797
Mean    :  16.89    Mean    : 79.360    Mean    :2002
3rd Qu.:  70.90    3rd Qu.:131.800    3rd Qu.:2540
Max.    : 112.80    Max.    :184.800    Max.    :3571
```

The aquifer data set contains piezometric head heights (in feet above sea level) for the Wolfcamp Aquifer in West Texas [(Cressie, 1989, p. 212), (Harper and Furr, 1986)]. The easting and northing columns of the data frame are locations. If you want to see the summary statistics for the head data only (excluding the location data), use the command:

```
> summary(aquifer$head)
 Min. 1st Qu. Median Mean 3rd Qu. Max.
 1024    1548    1797 2002    2540 3571
```

If you are only interested in the mean or median, use the functions mean or median. The large difference between the median and the mean of the aquifer head data indicates that the distribution may be skewed, or that there may be influential outliers.

A simple way to look at the one-dimensional distribution of your data is with a stem-and-leaf diagram as follows:

```
> stem(aquifer$head, twodig=T)
N = 85    Median = 1797    Quartiles = 1548, 2540
Decimal point is 2 places to the right of the colon
   10 : 24,30,38,89,92
   11 : 61
   12 : 31
   13 : 06,32,64,76,84,86
   14 : 02,08,15,37,64,66,76
   15 : 27,48,79,91
   16 : 06,11,38,74,80,82
   17 : 02,14,22,25,29,35,36,39,56,57,71,77,97
   18 : 05,06,28,65,68
   19 : 99
   20 : 03
   21 : 18,58
   22 : 00,38
   23 : 00,52,86
   24 : 00,32,55,68
   25 : 28,33,40,44,53,53,60,75,94
   26 : 46,48,50,91
   27 : 28,29,36,66,98
   28 : 11
   29 : 46
   30 :
   31 : 36
   32 :
   33 :
   34 : 90
   35 : 10,71
```

The stem-and-leaf diagram is a sideways histogram, but gives more information, as the individual values can be read directly from the output. There is one *leaf* for each data value, and the leaves are divided into ordered *stems* of equal bin size. The numbers to the left of the colons above are the first two digits of the head data, and the numbers to the right of the colon are the last digits. The optional argument `twodig=T` has been specified so that all four digits are displayed. For example, the row which begins with `21:` indicates that there are two data values between 2096 and 2195; they are 2118 and 2158.

The distribution of the aquifer data appears to be bimodal, except for some large outliers (3490, 3510, 3571), and an extra group of low values (1024, 1030, 1038, 1089, 1092). In subsequent analysis, these specific features of the data should be taken into account.

3.1.2 Plots and Graphics

The most well-known plot of the frequency distribution of a data set is the histogram. Produce the plots in figure 3.1 as follows:

```
> motif()
> par(mfrow=c(2,1))
> hist(aquifer$head)
> hist(aquifer$head, nclass=20, xlim=c(1000,4000))
```

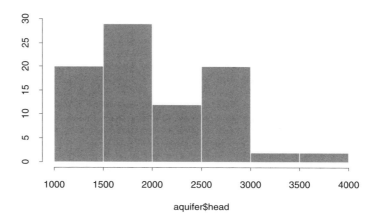

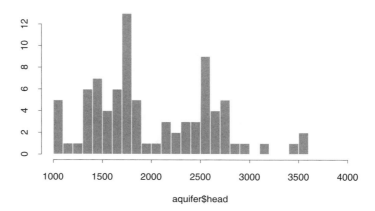

FIGURE 3.1. Histograms of Wolfcamp aquifer data.

To produce plots in S-Plus you must first open a graphics device—we have chosen `motif` in this example. Other choices include `openlook`, `trellis.device`, or `win.graph` for Windows applications.

The `par` function is used to change default parameters of the graphics device. In this case, we wanted to see two plots at once—row by row in a 2×1 arrangement on the page, so we used the argument `mfrow`.

The first histogram in figure 3.1 used the default number of classes (bars), which are too wide to show the bimodality and outliers evident in the stem-and-leaf plot. For the second histogram, we added the optional argument `nclass=20` to increase the number of bars, as well as another optional argument, `xlim = c(1000,4000)` which ensures the same data spread in the x-axis thereby allowing us to better compare the two histograms. (We could have also used the argument `ylim=` to affect the y-axis.) This plot clearly shows the bimodal distribution of the aquifer head data, as well as the potential outliers or extreme high values.

If there is a stratifying variable in your data set, you can use the `histogram` function from the Trellis library to display a panel of histograms for each strata. To try this feature, first break the aquifer head data into four groups based on the natural breaks observed in the frequency distribution.

```
> head.groups <- cut(aquifer$head,
+       breaks=c(0,1100,2000,3000,3600))
> trellis.device(color=F)
> histogram( ~aquifer$head | head.groups)
```

Figure 3.2 shows the histogram for each grouping. The graphics window `trellis.device` is the best choice for Trellis graphics. If you do not have a color screen, the argument `color=F` should be used. See the S-Plus *Trellis Displays User's Manual* for more information on Trellis graphics.

We have already observed bimodality in the aquifer data. A further check on its non-normality can be accomplished using a quantile plot of the head data vs. the standard normal distribution:

```
> qqnorm(aquifer$head)
> qqline(aquifer$head)
```

If the data came from an underlying normal distribution, the points produced by `qqnorm` and displayed in figure 3.3 should approximate a straight line. The function `qqline` can be used to plot a straight line fit to the `qqnorm` points for comparison. It is clear from the sigmoidal pattern of the points that the head data are not likely to be normally distributed, as we have already observed.

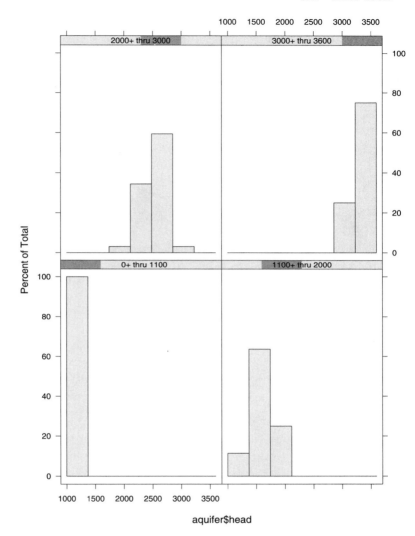

FIGURE 3.2. Histograms of aquifer heads by strata using Trellis graphics.

After a look at distributional data characteristics, we would like to add the spatial information to our explorations. A simple way to start is to map the data collection locations. To produce the scatter plot in figure 3.4 for the aquifer data:

```
> scaled.plot(x=aquifer$easting, y=aquifer$northing)
```

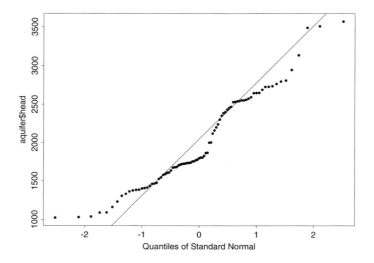

FIGURE 3.3. Plot of aquifer data vs. standard normal quantiles.

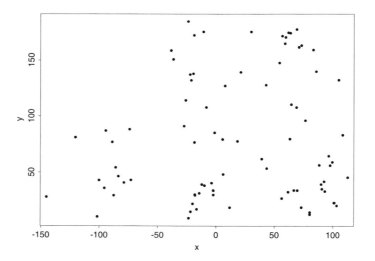

FIGURE 3.4. Data collection locations for Wolfcamp.

The function `scaled.plot` sets both axes to the same scale, which is essential when exploring spatial data.

To get an idea of the spatial distribution of the data groupings observed in the histogram of the aquifer head data (figure 3.1), we can use different plotting symbols for the groupings as follows:

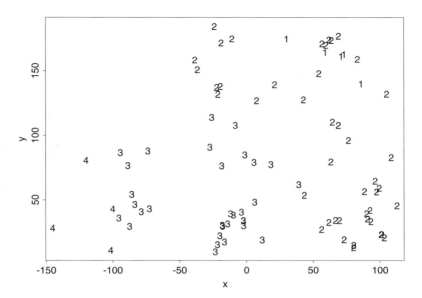

FIGURE 3.5. Locations for Wolfcamp, showing four groupings of data.

```
> scaled.plot(aquifer$easting, aquifer$northing, type="n")
> text(aquifer$easting, aquifer$northing,
+ labels=head.groups)
```

The plot argument type="n" plots a blank page with the correct axis scaling—this is a necessary preparation for using text which will then add textual labels corresponding to each point. Figure 3.5 shows an apparent *spatial trend*, with the highest values (group 4) in the lower left corner of the plot.

The trend observed in figure 3.5 can be confirmed with the two-dimensional trend visualization tools used to create figure 3.6. Produce these plots as follows:

```
> par(pty="s", mfrow=c(1,2))
> attach(aquifer)
> int.aq <- interp(x=easting, y=northing, z=head/1000)
> contour(int.aq)
> points(easting,northing)
> image(int.aq)
> detach("aquifer")
```

The par(pty="s") command creates a square plotting region, which is often desirable for displaying location data. The default plotting region in

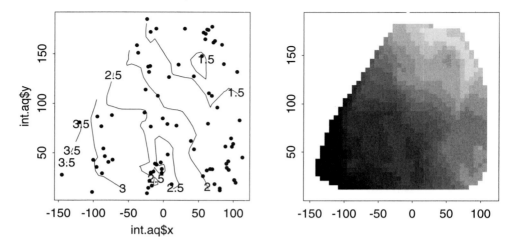

FIGURE 3.6. Two ways to visualize spatial trend in the Wolfcamp aquifer data:
a contour map (left); and a grayscale or image map (right).

S-PLUS is `pty="m"` which makes maximum use of the space allowed, but
may look distorted due to different x and y scaling. The function `attach`
adds the `aquifer` data frame to the second position in the search list,
allowing us to use the names of individual columns of the data frame in
subsequent calls to functions.

Note: It is recommended that you use `detach` to remove an attached
data frame from the search list when you are no longer using it.

The `interp` function interpolates head values on an equally spaced grid of
easting and northing values. The head values were divided by 1000 to make
the plots easier to read. The resulting list, with x and y coordinates and
a corresponding matrix of z values, is the data structure required to build
the contour plot and the grayscale plot. Both plots in figure 3.6 display the
spatial trend.

There are also several 3-dimensional plots which may be useful for
displaying trend. For the aquifer data:

```
> par(mfrow=c(1,1))
> persp(int.aq, zlab="Z/1000", eye=c(300,-1500,15))
> trellis.device(color=F)
> cloud(head ~ easting*northing, data=aquifer)
```

The 3-D perspective plot created by `persp` is displayed in figure 3.7. We
have used the optional argument `eye` to rotate the image. The 3-D point

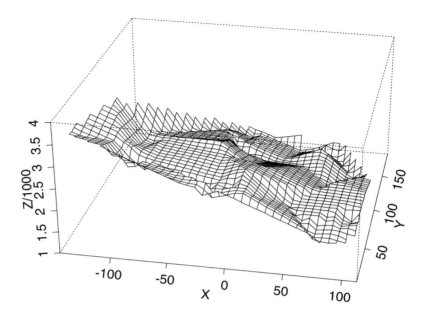

FIGURE 3.7. A 3-D perspective plot of the aquifer heads.

cloud displayed in figure 3.8 has been created with the `cloud` function from the Trellis library. The spatial trend is again observed in both 3-D figures.

3.1.3 Classification and Clustering Methods

Standard classification and clustering methods for multivariate data are available in S-PLUS. In this section we will briefly discuss regression trees and model-based clustering methods. Information on other multivariate methods such as hierarchical clustering (`hclust`), factor analysis (`factanal`), discriminant analysis (`discr`), and principal components analysis (`princomp`) can be found in help files for the individual functions, in the S-PLUS *Guide to Statistical and Mathematical Analysis*, and in other books listed in the bibliography; for example (Venables and Ripley, 1994).

Tree-based regression can be used as an exploratory technique for discovering structure in multivariate data. For spatial data analysis, it can aid in finding local neighborhoods where data are homogeneous within but with differences between the neighborhoods. The `tree` function in S-PLUS partitions a response variable based on a set of predictors. If we partition the aquifer head data by locations, we would expect to see the groupings we have already observed.

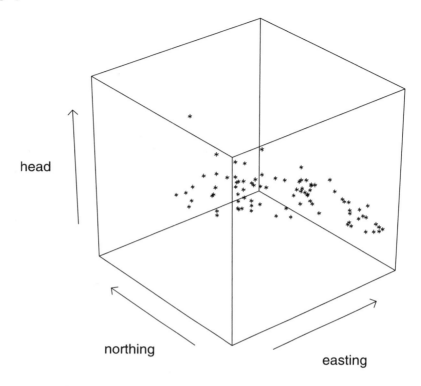

FIGURE 3.8. A 3-D point cloud of the aquifer heads.

```
> aqtree <- tree(head ~ easting + northing, data=aquifer,
+       mindev=.001)
> plot(aqtree)
```

The argument `mindev` sets the minimum deviance for each node. To start with a fairly large tree, we set the deviance relatively low. The `plot` command displays the structure of the tree (figure 3.9), with the default size of the branches determined in relation to the importance of the split.

Some of the lower branches in figure 3.9 do not seem very important. The S-Plus function `prune.tree` reduces the tree to an optimal size. First, choose an appropriate size by looking at the deviance plot:

```
> plot(prune.tree(aqtree))
```

The plot displayed in figure 3.10 shows that the deviance is not substantially reduced past the fifth node. Prune the tree to the most important five nodes and plot the result as follows:

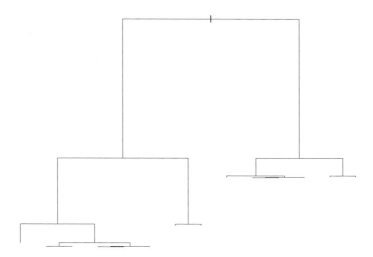

FIGURE 3.9. Regression tree for the aquifer data.

```
> aqtree2 <- prune.tree(aqtree, best=5)
> par(mfrow=c(1,2))
> plot(aqtree2, type="u")
> text(aqtree2, srt=90)
```

The left plot in figure 3.11 shows the pruned regression tree. Branches of uniform length were created using the argument `type="u"` inside the `plot` command. The generic `text` command used on a tree object prints the node information at each node. The optional argument `srt=90` rotates the text vertically so that the labels do not overlap. The number printed at each terminal node in the regression tree is the mean of the data which remains in that node after partitioning.

Check the spatial locations of the nodes of the regression tree as follows:

```
> par(pty="s")
> partition.tree(aqtree2)
> points(aquifer$easting, aquifer$northing)
```

The `partition.tree` function plots the spatial groupings produced by the `tree` function. This is the right plot in figure 3.11. We have added the points to help judge the sizes of the groups. The spatial trend observed in figure 3.6 is also apparent through these divisions.

There are other tree-related functions available in S-PLUS for shrinking (`shrink.tree`), browsing (`browser.tree`), and snipping (`snip.tree`) regression trees. See the S-PLUS *Reference Manual* and the S-PLUS *Guide*

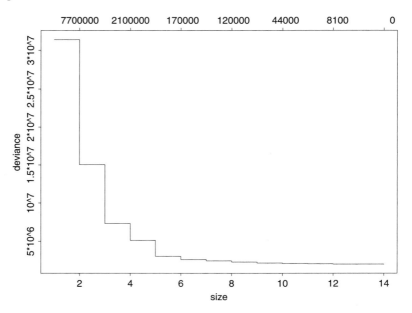

FIGURE 3.10. Plot of the reduction in deviance with increased size of the regression tree for the aquifer data.

to Statistical and Mathematical Analysis for further information on these topics.

Cluster analysis is another way of uncovering multivariate data structure. The S-PLUS function `mclust` performs *hierarchical agglomerative clustering* using any of six model-based or five heuristic criteria. For a list of the clustering criteria available when using `mclust`, see the S-PLUS *Guide to Statistical and Mathematical Analysis*. Hierarchical agglomerative clustering starts with a separate group for each object, then successively groups the objects into larger clusters based on some distance measure. Another method of clustering, *iterative relocation*, starts with an initial classification and attempts to iteratively improve on the fit. Iterative relocation clustering is available in S-PLUS using the function `kmeans`.

Perform EDA on the aquifer data using clustering techniques as follows:

```
> mclust.aq <- mclust(data.matrix(aquifer))
> xy <- plclust(mclust.aq$tree,labels=F)
> lab.aq <- as.character(aquifer$head)
> labclust(xy, lab.aq, cex=.75)
```

The `mclust` function requires data in the form of a matrix, so we first use the `data.matrix` function on the aquifer data frame. The tree produced

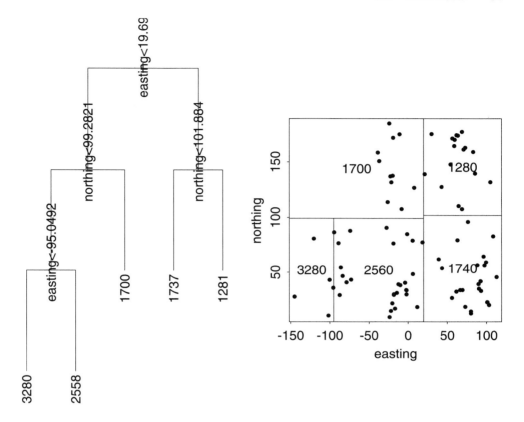

FIGURE 3.11. Pruned regression tree for the aquifer data with a partition plot.

by `plclust` in figure 3.12 shows the clusters in the aquifer data, formed using the default (S*) clustering criterion. We have labeled the clusters with the head values, using the function `labclust`. The optional argument `cex = .75` changes the text height to 75 percent of the default height.

 Warning: The clusters produced by `mclust` are dependent on the scale of the variables involved. Since the spatial units are arbitrary, this technique must be used with caution.

To see where the top five clusters produced by `mclust` are spatially located, plot the clusters as shown in figure 3.13:

```
> aquifer.5 <- mclass(mclust.aq, 5)
> scaled.plot(aquifer$easting, aquifer$northing, type="n")
> text(aquifer$easting, aquifer$northing, aquifer.5$class)
```

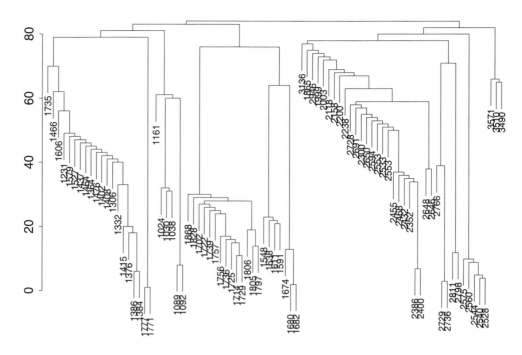

FIGURE 3.12. Tree produced by model-based clustering on the aquifer data.

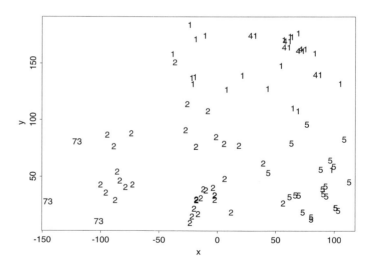

FIGURE 3.13. Map of Wolfcamp aquifer locations showing five clusters produced by model-based clustering.

The function `mclass` takes an object produced by `mclust` and returns a vector classifying the original data into clusters. By default, the clusters, shown in figure 3.13, are labeled by the first row number of the data frame that was included in each cluster. For example, the first element included in the cluster labeled by "73" is found in the 73rd row of the data frame. In figure 3.5, we plotted the spatial locations of the data groupings seen in the bimodal histogram. Compare this to figure 3.13. The groupings are similar, but adding the spatial dimension has added a possible cluster in the southeast corner (labeled by "5") which we had not noticed before.

3.2 Applications of EDA to Geostatistical Data

In this section, we will give specific examples of EDA for *geostatistical data*—data collected on a continuous spatial surface (see chapter 1 for a more precise definition). The `coal.ash` data frame is used in an example of EDA for data collected on an equally spaced grid of locations (section 3.2.1). The `scallops` data frame is used in the example of EDA for non-gridded data (section 3.2.2).

3.2.1 Example: Gridded Data

The `coal.ash` data come from the Pittsburgh coal seam on the Robena Mine Property in Greene County, Pennsylvania [(Cressie, 1993, p. 32), (Gomez and Hazen, 1970)]. The data frame contains 208 coal ash core samples collected on a grid given by x and y planar coordinates. Plot the grid locations displayed in figure 3.14 as follows:

```
> plot(coal.ash$x, coal.ash$y, type="n", lab=c(16,23,7))
> text(coal.ash$x, coal.ash$y,
+       format(round(coal.ash$coal,1)))
```

Figure 3.14 displays the data at their sampled locations. The graphical parameter `lab` within the `text` function specifies the number of tick intervals for each axis and the length of the axis labels. See the help file for `par` for more information on graphical parameters. The data were rounded to one decimal place using the functions `format` and `round`.

For questions about the distribution of coal ash, look at the descriptive statistics as follows:

```
> summary(coal.ash$coal)

  Min. 1st Qu. Median  Mean 3rd Qu.   Max.
     7    8.96  9.785 9.779   10.57  17.61
```

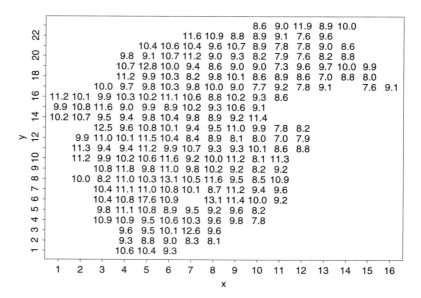

FIGURE 3.14. Core measurements (in percent coal ash) at sampled locations.

```
> stem(coal.ash$coal)

N = 208    Median = 9.785
Quartiles = 8.96, 10.575

Decimal point is at the colon

    7 : 003
    7 : 666788888999
    8 : 0111122222234
    8 : 56666666778888899999999
    9 : 000000001111111222223333333344444
    9 : 5555555666666666778888888888999999999
   10 : 000000001111111222222333334444444
   10 : 556666677777788888899999
   11 : 0000011112222223344
   11 : 5666679
   12 :
   12 : 568
   13 : 11

High: 17.61
```

A look at the summary statistics and the stem-and-leaf plot shows that the data are approximately normal, with some possible outliers. The output from the stem function, High: 17.61, tells us that at least one point, 17.61, was too high to print on the stem-and-leaf plot. (Try the optional argument fence to change the definition of outliers in the call to stem.)

So far, we have only looked for outliers and trends across the *global* surface. For spatial data analysis, we are also interested in observations which are atypical for a *local* neighborhood of points. Since this data is gridded, we can easily look for violations of local stationarity across rows and columns of the spatial gridded locations. One way to do this is using box-and-whisker plots as shown in figures 3.15 and 3.16. First, find and remove the suspected outlier so that it will not distort the scale of the figures:

```
> coal.ash[coal.ash$coal==max(coal.ash$coal),]
   x y  coal
50 5 6 17.61

> trellis.device(color=F)
> bwplot(y~coal, data=coal.ash, subset= -50,
+      main="Row Summaries")
> bwplot(x~coal, data=coal.ash, subset= -50,
+      main="Column Summaries")
```

We have used the bwplot command from the Trellis library of functions, separating the coal ash data first by row (figure 3.15), then by column (figure 3.16). Several outliers appear in both plots, indicating points which need to be more closely examined. There also may be a trend across columns.

Another way to look for local stationarity, suggested by Cressie (1993), is to plot the means and medians across rows and columns. The S-PLUS function tapply applies a function such as mean or median to a subset of a data frame. Produce the plots in figure 3.17 as follows:

1. Set up graphics device for four separate plots.

    ```
    > par(mfrow=c(2,2))
    ```

2. Plot locations without axes or labels.

    ```
    > plot(coal.ash$x, coal.ash$y, axes=F, xlab="",
    +     ylab="")
    ```

3. Plot row medians, then add row means to the plot.

Row Summaries

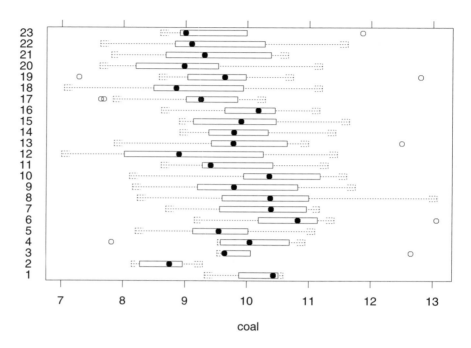

coal

FIGURE 3.15. Boxplots for coal ash data separated by row.

```
> plot(tapply(coal.ash$coal, coal.ash$y, median),
+       1:23, xlab="% coal ash", ylab="Rows",
+       xlim=c(8,12), pch="o")
> points(tapply(coal.ash$coal, coal.ash$y, mean),
+       1:23, pch="x")
```

4. Plot column medians, then add column means.

```
> plot(1:16, tapply(coal.ash$coal,coal.ash$x,median),
+       xlab="% coal ash", ylab="Columns",
+       ylim=c(7,11), pch="o")
> points(1:16, tapply(coal.ash$coal, coal.ash$x,
+       mean), pch="x")
```

5. Plot the legend.

```
> plot(coal.ash$x, coal.ash$y, axes=F, type="n",
+       xlab="", ylab="")
```

Column Summaries

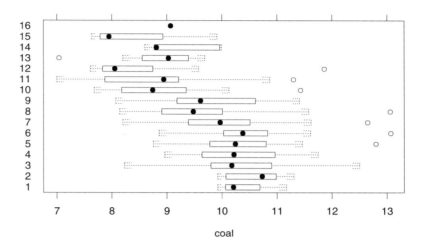

FIGURE 3.16. Boxplots for coal ash data separated by column.

```
> text(1,22, "o = Median % coal ash", adj=0)
> text(1,19, "x = Mean % coal ash", adj=0)
```

Figure 3.17 shows the apparent trend across the columns as seen in the box-and-whisker plot. Outliers or a skewed distribution within a row or column are indicated by large differences between the mean and the median.

Another way of detecting violations of local stationarity is to look at spatially lagged plots—bivariate scatter plots of each point versus its neighbor in a given direction. Produce these plots as follows:

1. Create a matrix of coal ash values with x as columns and y as rows.

```
> grid.mat <- tapply(coal.ash$coal,
+       list(factor(coal.ash$x), factor(coal.ash$y)),
+       function(x)x)
```

2. Plot columns 2 through 23 vs. columns 1 through 22, and identify interesting points.

```
> par(pty="s")
> plot(grid.mat[,-1],grid.mat[,-23],xlab="coal ash% in
+       column Z",ylab="coal ash% in column Z+1")
> identify(grid.mat[,-1], grid.mat[,-23],
+       label=grid.mat[,-1])
[1] 179 277  72  23
```

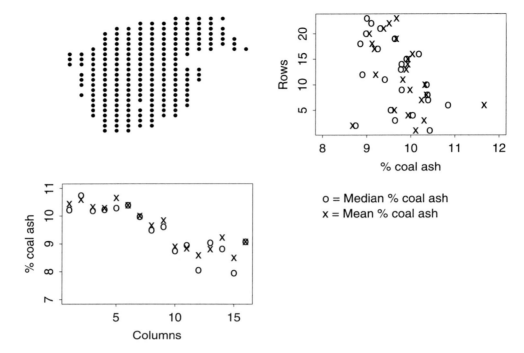

FIGURE 3.17. Data summaries for coal ash data across rows (top right) and columns (bottom left).

3. Plot rows 2 through 16 vs. rows 1 through 15, and identify interesting points.

```
> plot(grid.mat[-1,],grid.mat[-16,],
+       xlab="coal ash% in row Z",
+       ylab="coal ash% in row Z+1")
> identify(grid.mat[-1,], grid.mat[-16,],
+       label=grid.mat[-1,])
[1]   79 110
```

The left plot in figure 3.18 shows the neighbor plot across columns, and the right plot shows the neighbor plot across rows. We have used the S-PLUS function identify to label potential outliers. The previously detected outlier $z[5,6] = 17.61$ is apparent on both plots. The notation $z[x,y]$ indicates the value in the 5th column and the 6th row. Other points which seem atypical compared to neighboring points, according to Cressie (1993), include $z[7,3] = 12.65$, $z[8,6] = 13.06$, $z[6,8] = 13.07$, $z[3,13] = 12.5$, and $z[5,19] = 12.8$. This exploration has led to a set of unusual observations which should be re-evaluated when variogram models are attempted.

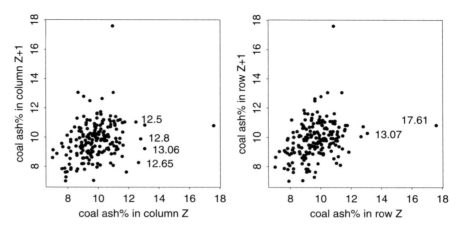

FIGURE 3.18. Bivariate scatter plot of each coal value versus its neighbor in the next column (left) or row (right).

3.2.2   Example: Non-gridded Data

The sample data set of scallop abundance in the Atlantic Ocean off the northeast U.S. coast is an example of a non-gridded geostatistical data. The data were collected by the Northeast Fisheries Science Center of the National Marine Fisheries Service, using a stratified random sampling scheme (Ecker and Heltshe, 1994). There are seven columns in the scallops data frame:

```
> summary(scallops)

       strata            sample             lat
6310    :24     Min.    :  1.0    Min.    :38.60
6270    :17     1st Qu.:106.8    1st Qu.:39.46
6230    :16     Median :147.0    Median :39.98
6340    :14     Mean    :131.8    Mean    :39.91
6300    :14     3rd Qu.:185.2    3rd Qu.:40.41
6260    :12     Max.    :224.0    Max.    :40.92
(Other):51
         long             tcatch            prerec
Min.    :-73.70    Min.    :   0.0    Min.    :   0.00
1st Qu.:-73.14    1st Qu.:   8.0    1st Qu.:   1.00
Median :-72.74    Median :  30.0    Median :   8.00
Mean    :-72.72    Mean    : 274.6    Mean    : 156.50
3rd Qu.:-72.31    3rd Qu.: 115.2    3rd Qu.:  48.25
Max.    :-71.52    Max.    :7084.0    Max.    :4487.00
```

```
        recruits
Min.   :    0.00
1st Qu.:    5.00
Median :   21.50
Mean   :  118.10
3rd Qu.:   73.75
Max.   : 2597.00
```

The number of recruits (`recruits`) and the number of pre-recruits (`prerec`) sum to the total catch (`tcatch`.)

An initial look at summary statistics suggests highly skewed data, since the mean and median are far apart. Perform a log transformation on the `tcatch` values to approximate normality:

```
> scallops[,"lgcatch"] <- log(scallops$tcatch+1)
> summary(scallops$lgcatch)
 Min. 1st Qu. Median  Mean 3rd Qu.  Max.
    0   2.197  3.434 3.483   4.756 8.866
```

Begin looking at the spatial character of the scallops data with a scatter plot including contours:

```
> library(maps)
Warning messages:
  The functions and datasets in library section maps are
        not supported by StatSci. in: library(maps)
> map("usa", xlim=c(-74,-71), ylim=c(38.2,41.5))
> points(scallops$long, scallops$lat, cex=.75)
> int.scp <- interp(scallops$long, scallops$lat,
+       scallops$lgcatch)
> contour(int.scp, add=T)
```

Figure 3.19 shows the catch locations in relation to the Atlantic coastline. The contours show a ridge of high catch values running northeast to southwest, with a sharp drop-off which might be expected as the ocean gets too deep for scallop survival. We should look at this spatial trend more carefully, since it will affect future modeling of this data.

One way to look for trends in non-gridded data is to model the data as a parametric trend surface using polynomials of longitude and latitude with the S-PLUS linear regression function lm. A better way for exploratory analysis is to model the data as a smooth function of the longitude and latitude. To accomplish this in S-PLUS, use the gam function to fit a generalized additive model (GAM) to the logged catch values, using longitude

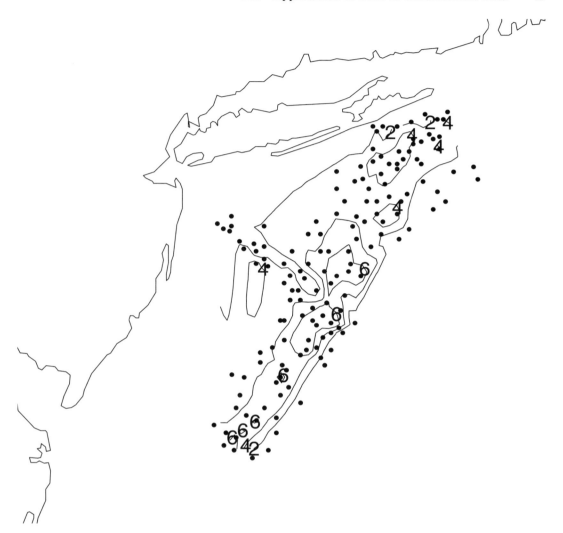

FIGURE 3.19. Map of scallop locations off the NE coast of the United States, including contour lines.

and latitude as the predictors. This model assumes independence, so the results must be used with caution.

```
> gam.scp <- gam(lgcatch ~ lo(long) + lo(lat),
+ data=scallops)
> par(mfrow=c(2,1))
> plot(gam.scp, residuals=T, rug=F)
```

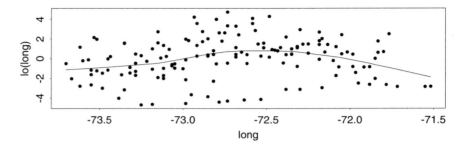

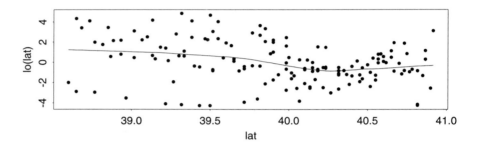

FIGURE 3.20. Plots of a smooth trend model across longitude (top) and latitude (bottom)

The use of `lo` specifies a loess, or local regression, fit for the **gam** function. The plots in figure 3.20 show no obvious trend running north-south or east-west. The fitted lines are not flat, but the residuals are too large for any strong conclusions.

 Warning: If you plot these loess fits without residuals, the trend may appear to be significant.

The spatial coordinate system given by longitude and latitude is arbitrary. In figure 3.19, we observed a trend running approximately perpendicular to the coastline. To look explicitly at this trend—or to remove it, we might try rotating the latitude and longitude axes. Write a function in S-PLUS to rotate the axes as follows:

```
> rotate.axis <- function(xy, theta)
+ {
+ # xy is an nx2 matrix of coordinates
+ # theta is the angle of rotation desired
+ # theta can be negative
+         pimult <- (theta * 2 * pi)/360
+         newx <- c(cos(pimult), sin(pimult))
+         newy <- c( - sin(pimult), cos(pimult))
```

```
+            XY <- as.matrix(xy) %*% cbind(newx, newy)
+   as.data.frame(XY)
+ }
```

After some testing with this function, an angle of 52 degrees—half way between 45 degrees and 60 degrees—seems appropriate. Plot the result of the 52 degree rotation (figure 3.21) and new **gam** fits (figure 3.22) as follows:

1. Rotate axes and plot the result.

```
> xy <- scallops[c("long","lat")]
> scall.rot <- cbind(rotate.axis(xy,52),
+       lgcatch=scallops$lgcatch)
> plot(scall.rot$newx,scall.rot$newy)
```

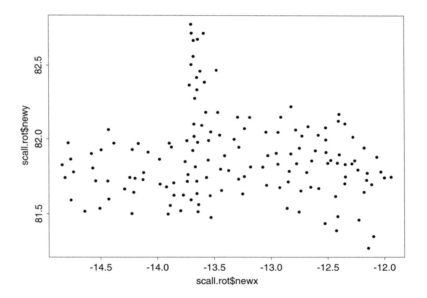

FIGURE 3.21. Plot of rotated scallop locations. North is now located at the top right corner of the plot.

2. Fit a GAM to the rotated data and plot the result.

```
> gam.scprot <- gam(lgcatch~lo(newx)+lo(newy),
+       data=scall.rot)
> par(mfrow=c(2,1))
> plot(gam.scprot, residuals=T)
```

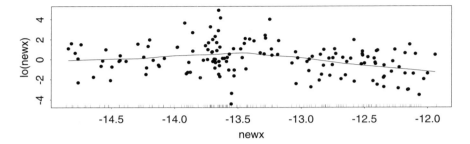

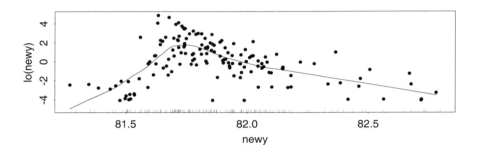

FIGURE 3.22. Plots of a smooth trend model across rotated longitude (top) and latitude (bottom).

The top plot in figure 3.22 shows the fitted smooth trend line running along the coastline from southwest to northeast. There doesn't appear to be a significant trend in this direction. The trend going from the ocean towards shore is approximated along the new latitude axis running southeast to northwest in the bottom plot in figure 3.22. This trend is shaped as could be expected, a sharp increasing gradient where the water becomes shallow enough for scallop survival, then a gentle decline towards shore.

The additive models used above are restricted to looking at trend in two directions. To fit a model to the whole trend surface, use the function `loess` as follows:

```
> loess.scp <- loess(lgcatch ~ newx*newy, data=scall.rot,
+       normalize=F, span=.25)
```

The function `loess` fits a local regression model to the data, returning an object of class `"loess"`. We have used the optional argument `normalize=F` to avoid normalizing the predictors, which are arbitrary locations in this case. We have set the `span` argument, the smoothing parameter, to .25, a larger value would produce a smoother fit while a smaller one would have the opposite effect. The residuals from this model will be used to form kriging predictions in chapter 4. For more information on loess models, see the S-PLUS *Guide to Statistical and Mathematical Analysis* or Cleveland, et al.(1992).

View the trend surface fit by the loess model above as follows:

1. Form predictions from the loess model on a grid covering the range of *newx* and *newy*.

```
> range(scall.rot$newx)
[1] -14.84007 -11.94120
> range(scall.rot$newy)
[1] 81.26967 82.77920
> lo.grid <- expand.grid(
+     newx = seq(-14.8,-11.9,length=50),
+     newy = seq(81.3,82.8,length=50))
> scp.pred <- predict(loess.scp, lo.grid)
```

The function `expand.grid` used above produces a 50×50 data frame containing all combinations of the *newx* and *newy* sequences. The function `predict.loess`, invoked by use of the generic `predict` function on a loess object, forms loess predictions on the grid.

2. Reduce the range of the predictions to match the non-rectangular area sampled for the scallops data.

```
> scall.chull <- chull(scall.rot$newx, scall.rot$newy)
> scall.poly <- list(x=scall.rot$newx[scall.chull],
+     y=scall.rot$newy[scall.chull])
> inside <- points.in.poly(lo.grid$newx, lo.grid$newy,
+     scall.poly)
> scp.pred[!inside] <- NA
```

The function `chull` returns the indices of the location values which form the convex hull surrounding the sampled sites. The function `points.in.poly` returns a logical vector indicating which points in the grid are contained in the polygon formed by `chull`.

3. View the predictions.

```
> persp(scp.pred)
```

The loess predictions are displayed in figure 3.23.

So far, our investigations into the `scallop` data have focused on global stationarity. What about local neighborhoods of points? In the gridded

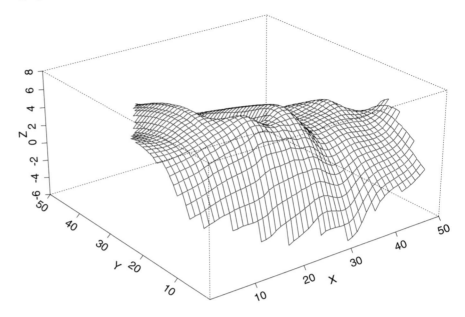

FIGURE 3.23. Perspective plot of the predictions of logged scallops data from a loess model.

data set, we looked at column and row summaries and spatial lagged plots. We can not do these plots on non-gridded data without grouping points to produce an artificial grid.

One way to look at local neighborhoods of points in S-PLUS is with the xyplot function from the Trellis library. Divide the scallops data into neighborhoods or *shingles* of points, then plot with xyplot as follows:

```
> y.shing <- equal.count(scall.rot$newy, number=6,
+       overlap=.25)
> xyplot(lgcatch~newx|y.shing, data=scall.rot)
```

The function equal.count creates shingles—six equally-sized groupings of the *newy* axis values which overlap by 25 percent. The plot in figure 3.24 shows the scatter of scallop catch values for different levels of our rotated axis *newy*. There appear to be a few local outliers (with values near zero) in the fourth (top left) and fifth (top middle) shingles.

The strata column in the scallops data frame contains locational strata created by the NMFS, which we may consider to be local neighborhoods.

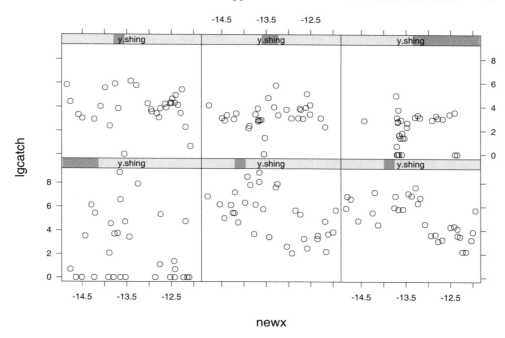

FIGURE 3.24. Plots of logged scallop catch at different levels of axis *newy*.

We can check the consistency of the catch values across strata with box-and-whisker plots. We are not concerned with "outliers" in strata having less than six observations, so remove these data before plotting:

```
> table(strata)
 6220 6230 6240 6250 6260 6270 6280 6290 6300
    8   16    5    3   12   17   10    5   14

 6310 6330 6340 6350
   24   10   14   10
> scp.y <- lgcatch[strata!=6240 & strata!=6250
+      & strata!=6290]
> scp.x <- strata[strata!=6240 & strata!=6250 &
+      strata!=6290]
> bwplot(scp.y ~ scp.x)
```

The `table` function counts how many observations are included in each stratum. Figure 3.25 reveals two potential outliers in stratum 6300: sample numbers 165 and 167 (rows 92 and 94).

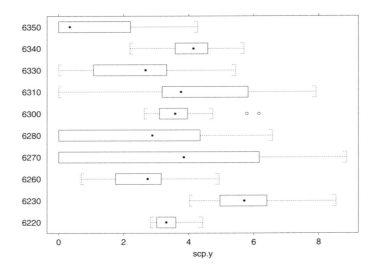

FIGURE 3.25. Box-and-whisker plots of logged scallop catch by strata.

3.3 Applications of EDA to Lattice Data

The sample data frame `sids` contains spatial data collected on a *lattice*. See chapter 1 for a definition of lattice data. The collection points are counties in the state of North Carolina, and the data are the rates of death from Sudden Infant Death Syndrome (SIDS) for the years 1974–1978 (Cressie and Chan, 1989). The components of the SIDS data frame are:

```
> names(sids)
[1] "id"       "easting"    "northing"
[4] "sid"      "births"     "nwbirths"
[7] "group"    "sid.ft"     "nwbirths.ft"
```

Data for the years 1979–1984 are also available in `sids2`. See the `sids` help file for explanations of the individual variables.

To form a spatial lattice, you must have data locations and *neighborhood* information. The locations for the SIDS data are stored in `easting` and `northing`. Neighborhood information is typically stored in a neighbor matrix, where two regions i and j are neighbors if the i,jth element of the neighbor matrix is non-zero. In S+SPATIALSTATS, neighbor information is stored in an object of class `"spatial.neighbor"`, a sparse matrix representation of the neighbor matrix. The S-PLUS object `sids.neighbor` contains the neighbor information for the SIDS data:

```
> sids.neighbor[1:15,]
Total number of spatial units =  100
(Matrix was NOT defined as symmetric)
    row.id col.id     weights matrix
  2      1     17 0.04351368      1
  3      1     19 0.04862620      1
  4      1     32 0.10268062      1
  5      1     41 0.20782813      1
  6      1     68 0.11500900      1
  8      2     14 0.17520402      1
  9      2     18 0.27140700      1
 10      2     49 0.20882988      1
 11      2     97 0.22700297      1
 13      3      5 0.16797229      1
 14      3     86 0.25458569      1
 15      3     97 0.22660765      1
 17      4     62 0.06865403      1
 18      4     77 0.15401887      1
 19      4     84 0.09565681      1
```

The columns `row.id` and `col.id` represent pairs of neighbors, and they correspond to the row index and column index for non-zero cells in the neighbor matrix. For example, the neighbors of region 1 are regions 17, 19, 32, 41, and 68. The strength of the neighbor relationship between each pair of neighbors is indicated by the `weights` column. The value in row 1, column 17 of the weighted neighbor matrix would be 0.0435. See section 5.1 for more information on neighbor weights and spatial neighbors.

For the SIDS data, two counties are defined as neighbors if their county seats are within 30 miles of each other, following the convention of Cressie (1993). Plot the SIDS lattice as follows:

```
> attach(sids)
> plot(easting, northing)
> segments(easting[sids.neighbor$row.id],
+     northing[sids.neighbor$row.id],
+     easting[sids.neighbor$col.id],
+     northing[sids.neighbor$col.id])
> detach("sids")
```

The function `segments` was used to draw line segments between all neighbor pairs. The lattice in figure 3.26 reveals two regions with no neighbors, and several regions with only one or two neighbors.

The variable `sids$sid` contains discrete counts of SIDS deaths for each county in the period 1974–1978. To accurately examine the spatial compo-

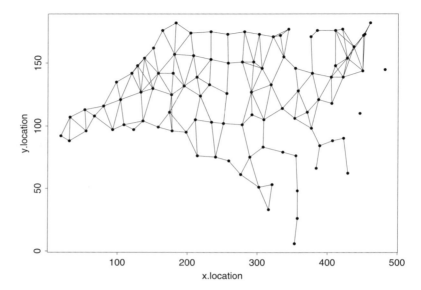

FIGURE 3.26. Plot of SIDS lattice.

nent of SIDS deaths in North Carolina, we must first correct for the total number of births in each county (`births`). We can create a standardized SIDS mortality rate, following the convention of Cressie (1993), as follows:

```
> sids.w <- 1000*(sids$sid+1)/sids$births
> hist(sids.w)
```

One is added to the SIDS count to distinguish among the counties with no SIDS deaths. The histogram displayed in figure 3.27 shows a skewed distribution with some potential high outliers.

To look for spatial patterns, we can use a map of the counties and color them based on the mortality rates. Assuming that the SIDS rates have Poisson distribution with a constant mean, create a probability map displaying the counties which lie in the tails of that distribution:

1. Find the common probability of SIDS, $\hat{p}$.

```
> attach(sids)
> sids.phat <- sum(sid)/sum(births)
```

2. Find the expected SIDS value for each county, $\hat{\lambda}$.

```
> sids.lambdahat <- births*sids.phat
```

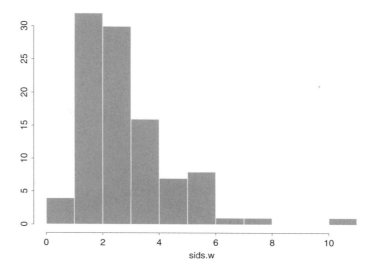

FIGURE 3.27. Histogram of SIDS mortality rates.

3. Calculate the cumulative probability for the actual SIDS rates, d_i, using the function ppois.

```
> sids.di <- ppois(sid, sids.lambdahat)
> detach("sids")
```

4. Create colors for the map: 1 for unusually high SIDS rates, and 4 for unusually low SIDS rates.

```
> sids.dcol <- rep(NA,100)
> sids.dcol[sids.di > .95] <- 1
> sids.dcol[sids.di < .05] <- 4
```

5. Map the result.

```
> library(maps)
Warning messages:
  The functions and datasets in library section maps
      are not supported by StatSci. in: library(maps)
> map("county", "north carolina", fill=T,
+     color=sids.dcol)
> map("county", "north carolina", add=T)
> legend(locator(1), legend=
+     c("Prob > .95","Prob < .05"), fill=c(1,4))
```

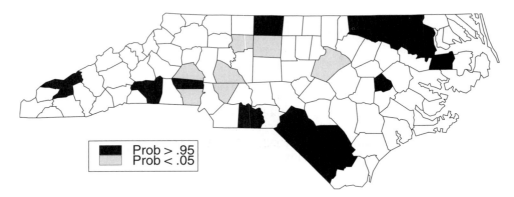

FIGURE 3.28. Probability map of SIDS mortality rates. Probabilities refer to the cumulative probabilities of the Poisson distribution.

The map in figure 3.28 displays two clusters of high values in the northeast and the south, and possibly a cluster of lower SIDS rates in the center of the state. This type of map is a useful exploratory technique, but the Poisson model ignores the likelihood of spatial autocorrelation between neighboring states.

Whether or not Poisson probabilities provide a good fit, count data is likely to have variance related to the mean, causing the map in figure 3.28 to be misleading. Counties with high birth rates will typically exhibit less variability in the number of SIDS deaths from year to year than counties with low birth rates. This problem, combined with the skewness seen in figure 3.27, is good cause for a data transformation. Assuming binomial distribution of the SIDS counts, a square-root transformation should remove the mean-variance dependence and skewness. But the addition of spatial dependence to a binomial model creates the necessity for a more severe transformation. Cressie and Read (1989) recommend the Freeman-Tukey square-root transformation:

$$Y_i = \sqrt{1000}\left(\sqrt{S_i/n_i} + \sqrt{(S_i+1)/n_i}\right)$$

where S_i is the SIDS count in county i and n_i is the birth count in county i. This transformation has been completed for you, and the data are stored in sids$sid.ft. Look at the stem-and-leaf plot for these transformed SIDS rates as follows:

```
> stem(sids$sid.ft)

N = 100    Median = 2.892998
Quartiles = 2.222682, 3.391945
```

```
Decimal point is at the colon

    0 : 9
    1 : 111244
    1 : 567789999
    2 : 0011111222334444
    2 : 5555566667777899999999
    3 : 000111122333333344444444
    3 : 5568999
    4 : 013344
    4 : 555557
    5 : 2
```

High: 6.283325

The stem-and-leaf plot of the transformed data above appears more symmetric, and only one outlier remains, county number 4. The variance is still dependent on the number of births, but Cressie and Read (1989) show that the variance is stabilized by this transformation. Therefore $\sqrt{n_i} * Y_i$, where n_i is the number of births in county i, and Y_i is the transformed SIDS rate in county i, should have approximately equal variances. This information] will be useful for the spatial regression modeling to come in section 5.3.

In section 3.2.1, we used spatially lagged plots to look for outliers in a local neighborhood of points for gridded geostatistical data (see figure 3.18). This technique can also be applied to gridded (or regular) lattice data. For irregular lattices, this technique only works if you first force your uneven lattice onto a gridded surface as in Cressie (1993). Haining (1990) suggests an alternative: plot each region versus the average of its neighbors.

Create a vector of neighbor averages for the transformed SIDS rates as follows:

1. Use the function `tabulate` to count the number of neighbors for each region.

   ```
   > sidstable <- tabulate(sids.neighbor$row.id)
   > sidstable
     [1] 5 4 3 4 3 6 3 4 2 1 6 6 4 7 4 1 4 6 7 2 8 3 3 2
    [25] 2 3 2 0 4 5 2 6 5 5 5 4 3 4 5 4 4 2 5 5 4 5 4 0
    [49] 7 4 4 3 3 4 5 7 4 3 6 4 5 4 4 3 2 2 1 5 2 5 2 7
    [73] 4 5 2 3 2 4 3 5 5 2 3 4 3 4 4 4 5 3 3 5 3 6 4 4
    [97] 7 5 5 5
   ```

2. Create a new spatial neighbor, with weights equal to the inverse of the number of neighbors for each region in `row.id`.

```
> sids.nhbr <- sids.neighbor
> sids.nhbr$weights <- 1/sidstable[sids.nhbr$row.id]
```

3. Create the vector of neighbor averages using `spatial.multiply`.

```
> sids.Ny <- spatial.multiply(sids.nhbr, sids$sid.ft)
```

The function `spatial.multiply` performs sparse matrix multiplication; it multiplies the neighbor weights for each region (row) times the appropriate SIDS rate in `sids$sid.ft`, then sums this to get the arithmetic average. The object `sids.Ny` is a (100×1) matrix with rows representing the neighbor averages for each county.

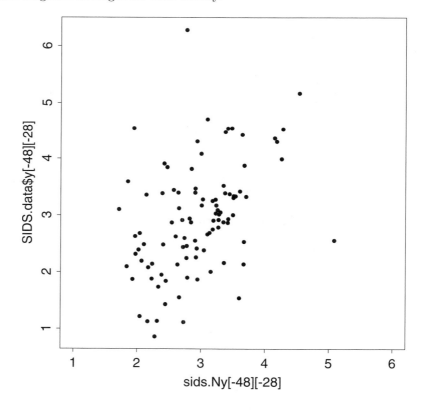

FIGURE 3.29. Scatter plot showing the average transformed SIDS rates of neighbors plotted against each county's transformed SIDS rate.

Before creating the plot, notice that there are two zero elements of `sidstable` (shown above), indicating two counties in the data set which have no neighbors. The corresponding zero values of `sids.Ny` should not be plotted:

```
> par(pty="s")
> plot(sids.Ny[-48][-28], sids$sid.ft[-48][-28],
+      xlim=c(1,6))
```

The only obvious outliers in figure 3.29 are county 4 and a neighbor average which contains county 4. At this point, it seems reasonable to eliminate county 4 from further analysis of this data set.

To continue our exploration of large-scale trend, we look for trend perpendicular to the x and y axes, as we did with non-gridded geostatistical data in section 3.2.2. Again, this model assumes independence, and will only be used in an exploratory sense.

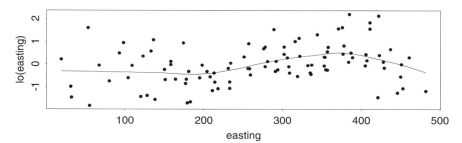

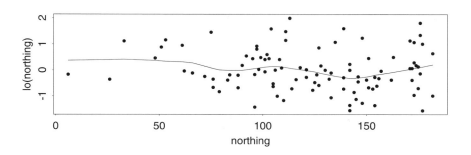

FIGURE 3.30. Plots of a smooth SIDS trend model across location axes.

```
> gam.sids <- gam(sid.ft~lo(easting) + lo(northing),
+      data=sids[-4,])
> par(mfrow=c(2,1))
> plot(gam.sids, residuals=T, rug=F)
```

The plots in figure 3.30 show the general trends that we observed on the map in figure 3.28. However, the relationships are not linear, and the residuals are quite large. Therefore, **easting** and **northing** may not be sufficient linear predictors of trend for the SIDS data.

Past analyses of this data set suggest that race may be an important co-variate for SIDS deaths [(Cressie, 1993, p. 550); (Cressie and Read, 1985); (Cressie and Chan, 1989); (Symons et al., 1983)]. The Freeman-Tukey transformed rates of non-white births in each county for the period 1974–1978 are contained in the variable sids$nwbirths.ft. Plot this rate versus the transformed SIDS rates to see the relationship:

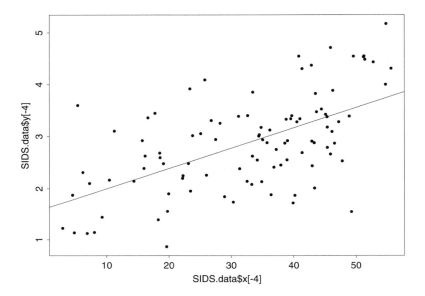

FIGURE 3.31. Scatter plot showing the transformed non-white birth rates plotted against the transformed SIDS rates.

```
> plot(sids$nwbirths.ft[-4], sids$sid.ft[-4])
> abline(lm(sids$sid.ft[-4] ~ sids$nwbirths.ft[-4]))
```

The outlier, county 4, was not included in figure 3.31. This plot shows an approximately linear relationship between the non-white birth rate and the SIDS rate. The line added using the function abline is the linear least squares regression line, found using the function lm. If non-white birth rate and county location are correlated, using non-white birth rate as a linear predictor in a trend model may account for the trend in location seen in figure 3.30. Plot easting versus non-white birth rates to see if they are positively associated:

```
> attach(sids)
> plot(easting[-4], nwbirths.ft[-4])
> abline(lm(nwbirths.ft[-4] ~ easting[-4]))
> detach("sids")
```

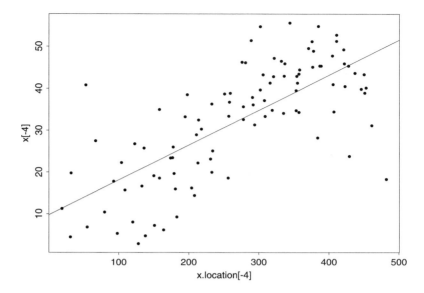

FIGURE 3.32. Scatter plot showing E–W location plotted against the transformed non-white birth rates.

The plot in figure 3.32 indicates that there may be a positive relationship between non-white birth rate and E–W location, although it does not appear to be linear.

Non-white birth rate is one possible covariate to include in a linear model of the SIDS rates in North Carolina. Obviously many other variables should be considered before a final model is chosen.

3.4 Applications of EDA to Point Patterns

A mapped spatial point pattern consists of a set of points or events located in a region of space. The data locations or points might be randomly located, tending to cluster in groups, or regularly located. A typical data analysis sequence begins with a test for *complete spatial randomness* (CSR), followed by an attempt to model any lack of spatial randomness. In this section, we will show how to begin exploring spatial point patterns. More formal checks for CSR and modeling techniques for spatial point patterns are described in chapter 6.

In S+SPATIALSTATS, locational data to be analyzed as spatial point patterns are stored in objects of class `"spp"`. Two columns in a point pattern object (typically the first two) are the locations of the points, and more columns may or may not be present. These objects can be created with data you scan into S-PLUS, with an existing S-PLUS data frame, or from two location vectors as follows:

```
> pines <- spp(matrix(scan("pine.dat"), byrow=T, ncol=2))

> bramble.spp <- as.spp(bramble)

> random.spp <- spp(x=runif(100), y=runif(100))
```

For this example, **pine.dat** is an ASCII file (not included with S+SPA-TIALSTATS) with data organized as x and y pairs, and **bramble** is a data frame in S-PLUS with the two first columns indicating data locations. The function **runif** generates a random sample from the uniform distribution on $[0, 1]$. The three results, **pines**, **bramble.spp**, and **random.spp** are objects of class **"spp"**.

The data frame **bramble**, included as part of S+SPATIALSTATS, contains the locations of 359 newly emergent bramble canes in a 9×9 meter plot [(Diggle, 1983, p. 83), (Hutchings, 1979)].

The locations have been scaled so that the data now reside in the unit square. Begin a spatial exploration of the bramble data by examining a simple spatial plot:

```
> is.spp(bramble.spp)
[1] T
> par(pty="s")
> plot(bramble.spp)
```

The function **is.spp** assures us that **bramble.spp** is a point pattern object. The plot method for spatial point patterns uses geometrically accurate axis scaling. The plot in figure 3.33 suggests clustering in the bramble data. A visual examination is not conclusive, however, since the human eye is apt to see patterns when there are none.

One aspect of spatial point patterns which affects analysis is the size and shape of the boundary region containing the locations. The analytical tools in S+SPATIALSTATS use rectangles for the boundary of spatial point patterns. If not supplied in the call to **spp**, this defaults to the bounding box found by the function **bbox**:

```
> bbox(bramble.spp)
$x:
[1] 0.026 0.026 0.997 0.997
```

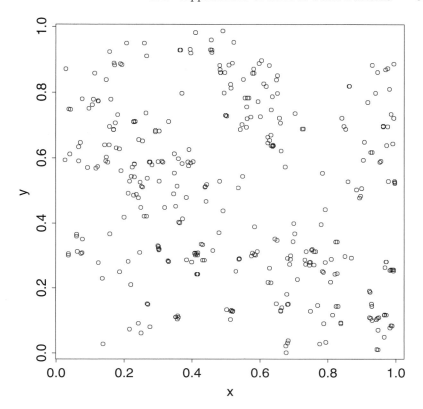

FIGURE 3.33. Scatter plot of bramble cane data.

```
$y:
[1] 0.987 0.001 0.001 0.987
```

This list is stored as the **boundary** attribute of an object of class **"spp"**:

```
> attributes(bramble.spp)$boundary
$x:
[1] 0.026 0.026 0.997 0.997

$y:
[1] 0.987 0.001 0.001 0.987
```

The bramble data were collected from a square plot which has been scaled to the unit square. The boundary can be redefined as the unit square as follows:

```
> bramble.spp <- spp(bramble, boundary=bbox(x=c(0,1),
+     y=c(0,1)))
```

In other cases, the boundary may be any convex polygon. This is represented in S+SPATIALSTATS as a list of the x and y vertex coordinates. For example, to find the minimum convex polygon enclosing all the points, the convex hull, use the function `chull` as follows:

```
> hull <- chull(bramble.spp)
> hull
 [1]   1  72 175 119 165 305 308 117 290 283 235
[12] 48   3  32
> bramb.poly <- list(x=bramble$x[hull],y=bramble$y[hull])
```

This polygon could then be used as the **boundary** argument for the function **spp**. Plot the bramble data with the two boundaries for comparison as follows:

```
> plot(bramble.spp, boundary=T)
> polygon(bramb.poly, density=0)
```

The dotted lines in figure 3.34 is the unit square boundary, and the solid line is the convex hull plotted with the function **polygon**. If you wish to expand the boundary by a fraction so that all of the points are completely within it, use the function **poly.expand**.

Next, we can estimate the overall *intensity*, or points per unit area, for the hull in figure 3.34. Simply divide the total number of points by the area of the polygon, found by using the function **poly.area**:

```
> poly.area(bramb.poly)
[1] 0.8789835
> 359/.879
[1] 408.4187
```

If the convex hull is the correct boundary, the estimated intensity is 408.4 events per unit area. This estimate for intensity assumes that the underlying process is stationary; the intensity is constant within the boundary region.

⟹   **Hint:** The intensity can also be estimated using the S+SPATIALSTATS function **intensity**, which will be introduced in chapter 6.

3.5 Hexagonal Binning

Hexagonal binning is a data grouping or reduction method typically employed on large data sets to clarify spatial structure. It can be thought of as partitioning a scatter plot into larger units to reduce dimensionality, while

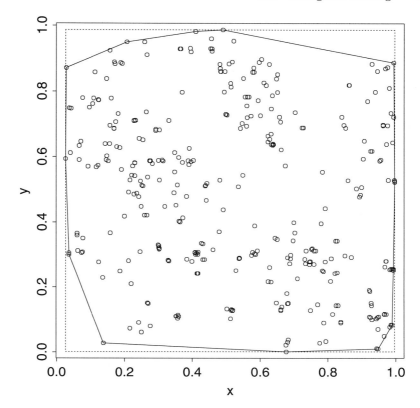

FIGURE 3.34. Scatter plot of bramble cane data showing the boundary and the convex hull.

maintaining a measure of data density. The groups or bins are used to make hexagon mosaic maps colored or sized according to density. Rectangular or square grids are often used in this context for image-processing applications, as seen in the grayscale, contour, and perspective maps in figures 3.6 and 3.7. However, hexagons are preferable for visual appeal and representational accuracy (Carr et al., 1992). Hexagonal binning might also be used to group geostatistical data into a lattice for use in spatial regression modeling.

The data frame `quakes.bay` contains the locations of earthquakes in the San Francisco Bay Area for 1962–1981. In S+SPATIALSTATS hexagonal bins are maintained in an object of class `"hexbin"`. Use the function `hexbin` to create the hexbin object for the earthquake data as follows:

```
> quakes.bin <- hexbin(quakes.bay$longitude,
+     quakes.bay$latitude)
```

```
> summary(quakes.bin)
Call:
hexbin(x = quakes.bay$longitude, y = quakes.bay$latitude)
Total Grid Extent:   36 by 31
        cell             count            xcenter
 Min.   :  17.0   Min.   :  1.000   Min.   :-123.3
 1st Qu.: 239.0   1st Qu.:  1.000   1st Qu.:-122.0
 Median : 419.0   Median :  3.000   Median :-121.6
 Mean   : 467.9   Mean   :  7.505   Mean   :-121.5
 3rd Qu.: 696.0   3rd Qu.:  5.000   3rd Qu.:-121.0
 Max.   :1091.0   Max.   :144.000   Max.   :-119.8

     ycenter
 Min.   :36.01
 1st Qu.:36.51
 Median :36.94
 Mean   :37.06
 3rd Qu.:37.59
 Max.   :38.50
```

The summary function shows the four components of the hexbin object, and their distributions. The hexagon identified by cell contains count observations, and has center of mass at (xcenter, ycenter). The default settings for hexbin partition the range of x values into approximately 30 equal-sided hexagonal bins. The most useful bin size depends on the number of observations, and is best chosen iteratively. Plot the hexagonal bins as follows:

```
> trellis.device(color=F)
> at.quakes <- c(0,10,20,30,40,50,150)
> plot(quakes.bin, border=T, col.regions=80:15,
+        at=at.quakes)
```

The Trellis graphics device produces the best color and grayscale images for hexagonal binning. The default settings for plot.hexbin plot the hexagonal bins as a full tessellation, containing equally sized hexagons with color corresponding to grouped bin counts. By default, the groups are equal in range. Since the distribution of quakes.bin$count (shown by the summary output above) is skewed, we have chosen the groups formed by at.quakes. The plot in figure 3.35 shows the ridge of frequent earthquakes along the San Andreas Fault.

Besides the default grayscale style used for figure 3.35, there are four other plot styles available which plot the hexagons in varying sizes depending on cell density. Plot the earthquake hexbin object with differing sizes of hexagons as follows:

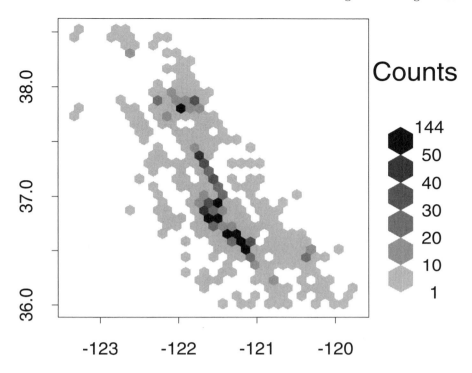

FIGURE 3.35. Plot of hexagonal bins of earthquakes in the San Francisco Bay Area from 1962 to 1981.

```
> plot(quakes.bin, style="centroids", cuts=6)
```

The "centroids" style shown in figure 3.36 scales the hexagon sizes by cell count, and plots them at the center of mass determined by xcenter and ycenter. The cuts = 6 argument yields six different hexagon sizes. There are two nested plot styles ("nested.lattice" and "nested.centroids", not shown) which provide visual depth when plotted on a color screen.

There are several large bins in figure 3.36 which we may want to examine more closely. The generic function identify can be used to interactively identify points on a hexagonal bin plot. Identify the two largest bins as follows:

```
> quake.par <- plot(quakes.bin, style="centroids", cuts=6)
> oldpar <- par(quake.par)
> identify(quakes.bin, use.pars = quake.par, offset=1)
[1] 114  79
> par(oldpar)
```

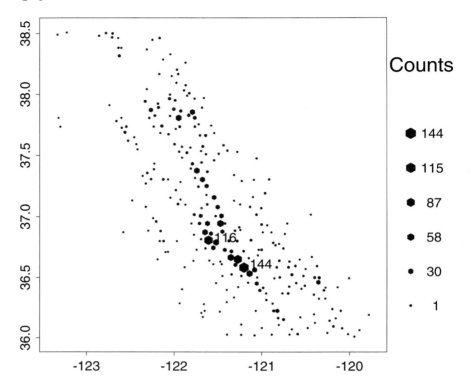

FIGURE 3.36. Plot of hexagonal bins of San Francisco Bay Area earthquakes using the "centroids" style.

First it is necessary to save the graphical parameters used to plot the hexagonal bin. After entering the `identify` command, use the cross-hairs to locate the point of interest on the graphics screen, and click the left mouse button. The count in the closest cell will appear on the graphics screen. We have used the optional argument `offset` to make the count easier to read (see figure 3.36). When you have identified both points, click the center or right mouse button, while keeping your pointer within the graphics window. The index of the points you have identified will appear on your command line, as above. Then use the `par` function to reset the graphics parameters.

The S+SPATIALSTATS function `rayplot` can be used to display the magnitudes of a variable of interest at spatial locations using directional rays. For smaller data sets, these rays or other types of symbols, can be plotted at each data location. However, when the number of sites is large, the magnitudes and trends are easier to visualize if the locations are first binned using `hexbin`. The following example uses the S-PLUS `ozone` data set:

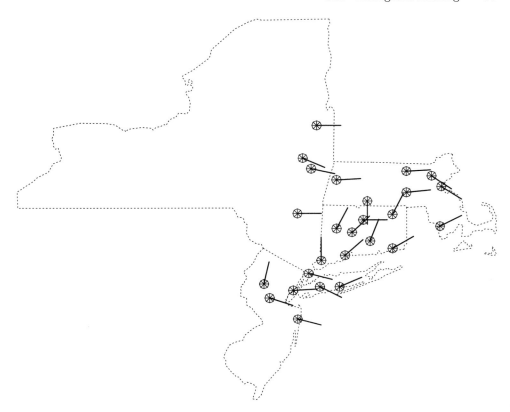

FIGURE 3.37. Rayplot of median ozone emissions.

1. Create a hexbin object for the ozone data, using eight bins in the x direction.

   ```
   > ozone.bin <- hexbin(ozone.xy$x, ozone.xy$y, xbins=8)
   ```

2. Map each (x, y) pair in the original data to a hexagonal cell using the function xy2cell.

   ```
   > ozone.cells <- xy2cell(ozone.xy$x, ozone.xy$y,
   +     xbins=8)
   ```

3. Use the function tapply to calculate the median for each cell, and use these values as angles for the rayplot.

   ```
   > ozone.angle <- tapply(ozone.median,ozone.cells,
   +     median)
   ```

```
> library(maps)
Warning messages:
  The functions and datasets in library section maps
are not supported by StatSci. in: library(maps)
> map(region=c("new york","new jersey","conn","mass"),
+      lty=2)
> rayplot(ozone.bin$xcenter,ozone.bin$ycenter,
+      ozone.angle)
```

The plot in figure 3.37 shows the median ozone emissions for the group of sites within each hexagonal bin. The ray is plotted at the center of each bin, and the medians are scaled so the rays follow an arc from $-\pi/2$ (lowest median) to $\pi/2$ (highest median). It appears that the highest emissions for the time period covered are in Connecticut. Additional attributes can be used with **rayplot** to add confidence intervals and/or a second variable to the plot. Also, the lengths and widths of the rays and the size of the base octagon can be changed. See the help file for more information on **rayplot**.

4

Analyzing Geostatistical Data

This chapter introduces functions available in S-PLUS and S+SPATIAL-
STATS for analyzing geostatistical data. Geostatistical data, also termed
random field data, consist of measurements taken at fixed locations. For
a complete description of geostatistical data see chapter 1. Specifically,
this chapter discusses methods related to variogram analysis and kriging.
Variogram estimation and kriging were originally introduced as geostatistical methods for use in mining applications. In recent years, these methods have been applied to many disciplines including meteorology, forestry,
agriculture, cartography, climatology, and fisheries.

In this chapter you will learn about the following topics:

- Estimating Variograms (section 4.1).

- Fitting Theoretical Variogram Models (section 4.2).

- Performing Ordinary and Universal Kriging (section 4.3).

- Simulating Geostatistical Data (section 4.4).

4.1 Variogram Estimation

Geostatistical data typically exhibit small-scale variation that may be modeled as spatial autocorrelation and incorporated into estimation procedures.
The variogram provides a measure of spatial correlation by describing how
sample data are related with distance and direction. In general, two closely

neighboring data are more likely to have similar values than two data farther apart. Other distance-based measures of spatial correlation include the
correlogram and covariogram functions.

In this section we continue exploratory data analysis for geostatistical data
by looking at tools that can be used to generate empirical variograms. Variogram estimation is exploratory and is often a multi-step process; a final
variogram model is constructed from knowledge of the underlying processes
affecting (generating) the data and by customization of the available tools.

S+SPATIALSTATS provides a variety of functions for use in variogram estimation. The subsections below describe tools for calculating empirical variograms, generating variogram clouds, detecting and removing trend, and
exploring and correcting for anisotropy. Variogram analysis is not necessarily performed following this linear sequence of topics; the final description
will likely be based on an iterative sequence of analyses using a combination of the possible tools. Your starting point for variogram analysis and
the order in which you choose to proceed, will depend on how much you
already know about your data. For example, if you know that your data
contains a trend, then your analysis might begin by modeling the trend
(see section 4.1.3).

4.1.1   The Empirical Variogram

The empirical variogram provides a description of how the data are related
(correlated) with distance. The semivariogram function, $\gamma(h)$, was originally defined by Matheron (1963) as half the average squared difference
between points separated by a distance h. The semivariogram is calculated
as

$$\gamma(h) = \frac{1}{2|N(h)|} \sum_{N(h)} (z_i - z_j)^2$$

where $N(h)$ is the set of all pairwise Euclidean distances $i - j = h$, $|N(h)|$ is
the number of distinct pairs in $N(h)$, and z_i and z_j are data values at spatial
locations i and j, respectively. In this formulation, h represents a distance
measure with magnitude only. Sometimes, it might be desirable to consider
direction in addition to distance. In such cases, h will be represented as the
vector $\mathbf{h}$, having both magnitude and direction.

Note: The terms semivariogram and variogram are often used inter-
changeably. By definition, $\gamma(h)$ is the semivariogram and the variogram
is $2\gamma(h)$. For conciseness, however, this manual will refer to $\gamma(h)$ as the
variogram.

The main goal of a variogram analysis is to construct a variogram that
best estimates the autocorrelation structure of the underlying stochastic

process. Most variograms are defined through several parameters; namely, the *nugget effect*, *sill*, and *range*. These parameters are depicted on the generic variogram shown in figure 4.1 and are defined as follows:

- *nugget effect*—represents micro-scale variation or measurement error. It is estimated from the empirical variogram as the value of $\gamma(h)$ for $h = 0$.

- *sill*—the $\lim_{h \to \infty} \gamma(h)$ representing the variance of the random field.

- *range*—the distance (if any) at which data are no longer autocorrelated.

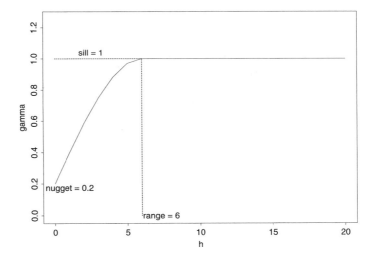

FIGURE 4.1. A generic variogram showing the *sill*, and *range* parameters along with a *nugget effect*.

Construction of a variogram requires consideration of the following:

- an appropriate *lag increment* for h;

- a *tolerance* for the lag increment; and

- the *number of lags* over which the variogram will be calculated.

The *lag increment* defines the distances at which the variogram is calculated. The *tolerance* establishes distance bins for the lag increments, to accommodate unevenly spaced observations. The *number of lags* in conjunction with the size of the lag increment will define the total distance over which a variogram is calculated.

There are two practical rules (Journel and Huijbregts, 1978) that should be considered when making your choices for lag increment and number of lags:

1. The experimental variogram should only be considered for distances h for which the number of pairs is greater than 30.

2. The *distance of reliability* for an experimental variogram is $h < D/2$, where D is the maximum distance over the field of data.

Omnidirectional Variograms

One way to begin a variogram analysis is to compute the omnidirectional variogram. An omnidirectional variogram is computed without respect to direction; all possible directions are combined into a single variogram. Use the S+SPATIALSTATS function `variogram` to generate the onmidirectional variogram for the coal ash data :

```
> coal.var1 <- variogram(coal ~ loc(x,y), data=coal.ash)
```

The `variogram` function requires a formula defining the response in terms of the spatial locations. The `loc` function is used to incorporate the location variables as predictors, and to distinguish them from the usual S-PLUS model specification for predictor variables. The `variogram` function also has a number of arguments that may be set to customize the variogram. These include `lag`, `nlag`, `tol.lag`, `maxdist`, and `minpairs`. By default, `variogram` yields an omnidirectional variogram with `maxdist` equal to the distance of reliability. The number of lags, `nlag`, is set to 20. The lag increment, `lag`, is automatically calculated as `maxdist/nlag`. The lag tolerance, `tol.lag`, defaults to `lag/2`.

The `variogram` function returns an object of class `"variogram"`. The components of `coal.var1` are the `distance`, $\gamma(h)$ (`gamma`), the number of pairs (`np`), and the `azimuth`:

```
> coal.var1
    distance    gamma   np azimuth
  1 1.201634 1.202911  719       0
  2 2.000000 1.172307  331       0
  3 2.236068 1.321759  644       0
  4 3.036036 1.314383 1170       0
  5 3.605551 1.297816  545       0
  6 4.234139 1.398687 1518       0
  7 5.039712 1.531598 1142       0
  8 5.472889 1.537340  638       0
```

```
 9   6.063483 1.536710 1382        0
10   6.552482 1.624369  719        0
11   7.147503 1.478895 1307        0
12   7.894487 1.491406 1269        0
13   8.459312 1.654917  882        0
14   9.094676 1.714728 1012        0
15   9.624306 1.804034  685        0
16  10.174499 1.656996  995        0
17  10.843929 1.705829  682        0
18  11.400797 1.880180  860        0
```

The calculated distance is the average distance between all pairs of points in the given distance bin. The azimuth is the clockwise angle from north, in degrees, defining the direction in which the variogram is calculated. The default onmidirectional variogram is based on `azimuth = 0`. To view a summary of the call to `variogram`:

```
> summary(coal.var1)
Call:
variogram.formula(formula = coal ~ loc(x, y),
                  data = coal.ash)
        lag nlag  maxdist
  0.6041523   20 12.08305
```

```
       distance              gamma                   np
Min.    : 1.202   Min.    :1.172   Min.    : 331.0
1st Qu.: 3.763   1st Qu.:1.341   1st Qu.: 682.8
Median : 6.308   Median :1.534   Median : 871.0
Mean    : 6.338   Mean    :1.518   Mean    : 916.7
3rd Qu.: 8.936   3rd Qu.:1.656   3rd Qu.:1163.0
Max.    :11.400   Max.    :1.880   Max.    :1518.0

        azimuth
         0:18
```

The summary includes the call to `variogram`, calculated default values for some of the arguments, and descriptive statistics for the returned values.

To plot the omnidirectional variogram for the coal ash data, shown in figure 4.2, use the generic S-PLUS `plot` function as follows:

```
> trellis.device()
> plot(coal.var1)
```

The `plot` function uses the plotting method `plot.variogram` for objects of class `"variogram"`. For single variograms, a standard plot of gamma versus distance is produced. The axes are set to include the point (0,0). Optionally, the user can set the `xlim` or `ylim` arguments.

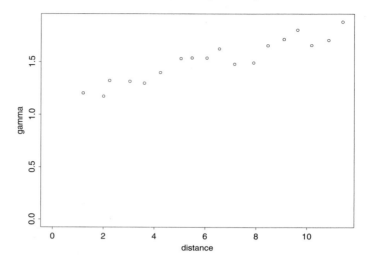

FIGURE 4.2. Omnidirectional empirical variogram for coal ash data.

The omnidirectional variogram for the coal ash data is generally increasing. This may indicate the presence of a large-scale trend or a nonstationary underlying stochastic process. Recall that the EDA of the coal ash data in section 3.2.1 showed an apparent trend in the east-west direction. The coal ash data set requires further analysis before settling on a variogram model.

S+SPATIALSTATS also provides the `covariogram` and `correlogram` functions . The covariogram is defined as:

$$cov(Z(i + h), Z(i)) = C(h), \quad \text{for all } i, i + h \in D$$

The correlogram, $\rho(h)$, is a ratio of covariances and is calculated as

$$\rho(h) = \frac{C(h)}{C(0)} = 1 - \frac{\gamma(h)}{C(0)}$$

where $C(h)$ is the covariance for pairs of points separated by Euclidean distances h (the covariogram), $C(0)$ is the finite variance of the random field, and $\gamma(h)$ is the corresponding variogram. These definitions are for the isotropic case, where h is a scalar; they can be extended to include the case where $\mathbf{h}$ is a vector with both magnitude and direction.

Create and plot the empirical covariogram and correlogram for the coal ash data, shown in figure 4.3, as follows:

```
# set up a 1x2 plotting window
> par(mfrow=c(1,2))
> coal.cov1 <- covariogram(coal ~ loc(x,y), data=coal.ash)
> plot(coal.cov1)
> coal.cor1 <- correlogram(coal ~ loc(x,y), data=coal.ash)
> plot(coal.cor1)
```

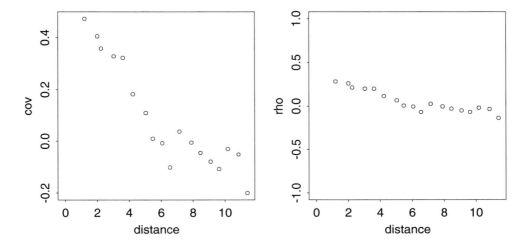

FIGURE 4.3. Omnidirectional empirical covariogram (left) and correlogram (right) for coal ash data.

The covariogram function returns an object of class "covariogram", while correlogram returns an object of class "correlogram". The objects returned from these functions are similar to that returned from a call to variogram. The distance, number of pairs (np), and azimuth are the same quantities. Either a $cov(h)$ or $\rho(h)$ (rho) value is returned, for each distance bin. As for a "variogram" object, a call to plot will invoke either the plot.covariogram or plot.correlogram methods. The lower limit of the y-axis of a covariogram defaults to the smaller of 0 and $min(cov)$. The y-axis limits for a correlogram default to $(-1, 1)$. The lower limit of the x-axis for both functions defaults to 0.

Directional Variograms

The variogram function can be used to produce *directional variograms*. In this case, $\gamma(h)$ is based on both the magnitude and direction of **h**. To generate a directional variogram, set the azimuth argument to the desired angle (relative to north). Multiple directional variograms can be computed by specifying a vector of azimuths. Calculate and plot some directional variograms for the coal ash data as follows:

```
> az <- c(0.0, 22.5, 45.0, 67.5, 90.0, 112.5)
> coal.var2 <- variogram(coal ~ loc(x,y), data=coal.ash,
+         azimuth=az, tol.azimuth=11.25)
> plot(coal.var2)
```

For multiple variograms, the method `plot.variogram` produces a multi-panel display that is drawn using the Trellis graphics `xyplot` function. Each panel contains a plot of gamma versus distance for a particular level of azimuth. The axes are set to include the point (0,0). The plots are best displayed on a device started with `trellis.device`.

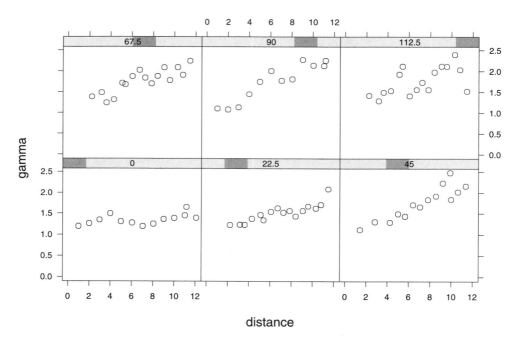

FIGURE 4.4. Directional empirical variograms for coal ash data.

The resulting directional variograms, plotted by `azimuth`, are shown in figure 4.4. The `tol.azimuth` argument is set to 11.25 so each directional variogram is based on all pairs of points that fall within the specified azimuth ±11.25 degrees.

The plots in figure 4.4 suggest that the north-south (`az=0`) and east-west (`az=90`) directional variograms are different. The north-south direction is basically flat, indicating little or no autocorrelation. The rest of the directions yield generally increasing variograms, which could be caused by the presence of trend and/or anisotropy, or some other form of nonstationary. See sections 4.1.3 and 4.1.4 for information on identifying and correcting for trend and anisotropy.

Robust Variogram Estimation

The variograms shown thus far have been generated using the classical formulation given by Matheron (1963). The `variogram` function has an option for calculating a robust estimator of the variogram developed by Cressie and Hawkins (1980). The robust estimation is based on the fourth power of the square root of absolute differences as follows:

$$\bar{\gamma}(h) = \frac{\left\{ \frac{1}{2|N(h)|} \sum_{N(h)} |z_i - z_j|^{1/2} \right\}^4}{0.457 + 0.494/|N(h)|}$$

where $N(h)$ is the set of all pairwise Euclidean distances $i - j = h$, $|N(h)|$ is the number of distinct pairs in $N(h)$, and z_i and z_j are data values at spatial locations i and j, respectively.

The advantage of the robust estimator is that the effect of outliers is reduced, without removing specific data points from a data set. To invoke the robust estimation, set the `method` argument to `"robust"`:

```
> coal.var3 <- variogram(coal ~ loc(x,y), data=coal.ash,
+       azimuth=az, tol.azimuth=11.25, method="robust")
> plot(coal.var3)
```

The directional variograms calculated using the robust estimator are shown in figure 4.5. The robust estimator appears to have smoothed out some of the variability that is present in the variograms (figure 4.4) based on the classical estimator.

4.1.2   Variogram Clouds

The *variogram cloud* is a diagnostic tool that can be used in conjunction with boxplots to look for potential outliers or trends, and to assess variability with increasing distance. Anomalies and nonhomogeneous areas can be detected by looking at short distances that yield high dissimilarities (large values of γ).

A variogram cloud is the distribution of the variance between all pairs of points at all possible distances h. The S+SPATIALSTATS `variogram.cloud` function allows you to specify a variance function of interest; most common

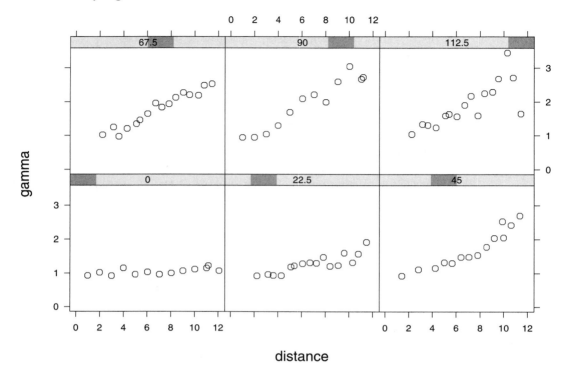

FIGURE 4.5. Robust directional empirical variograms for coal ash data.

are the squared-differences and the square-root-differences. The squared-differences cloud (the default) produces the distribution on which the classical variogram estimator is based, and results in a plot of $(Z_{i+h} - Z_i)^2/2$ versus h. The square-root-differences cloud yields a plot of $\sqrt{(|Z_{i+h} - Z_i|)}/2$ versus h.

Create and plot the default omnidirectional variogram cloud for the scallops data as follows:

```
> scallops.vcloud1 <- variogram.cloud(log(tcatch+1)
+        ~ loc(lat,long), data=scallops)
> # restore full screen plot
> par(mfrow=c(1,1))


> # restore maximum plotting area
> par(pty="m")
> plot(scallops.vcloud1)
```

The resulting squared-differences variogram cloud is shown in figure 4.6. The variogram cloud is extremely dense and shows the existence of many similar data values across all distances. The variability at small distances appears a bit less than that for larger distances.

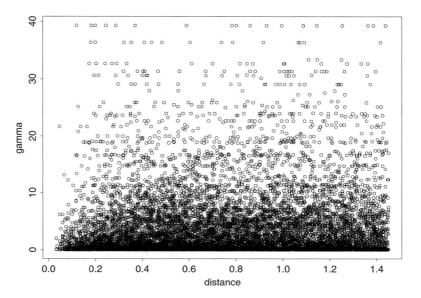

FIGURE 4.6. Squared-differences variogram cloud for scallops data.

An extremely dense variogram cloud may be difficult to interpret. We can reduce the density of the plot by reducing the maximum distance over which the variances are calculated. The optional argument `maxdist` defaults to the distance of reliability (1.5 for the scallops data). Reduce `maxdist` until a visible separation occurs as follows:

```
> par(mfrow=c(2,2))
> scallops.vcloud3 <- variogram.cloud(log(tcatch+1)
+       ~ loc(lat,long), data=scallops, maxdist=.5)
> plot(scallops.vcloud3)
> scallops.vcloud4 <- update(scallops.vcloud3,maxdist=.25)
> plot(scallops.vcloud4)
> scallops.vcloud5 <- update(scallops.vcloud3,
+       maxdist=.125)
> plot(scallops.vcloud5)
> scallops.vcloud6 <- update(scallops.vcloud3,
+      maxdist=.0625)
> plot(scallops.vcloud6)
```

The **update** function is used to create the successive variogram clouds displayed in figure 4.7; **update** adjusts the original call to **variogram.cloud** with the new value of **maxdist**. The **update** function can also be used to easily modify calls to **variogram**.

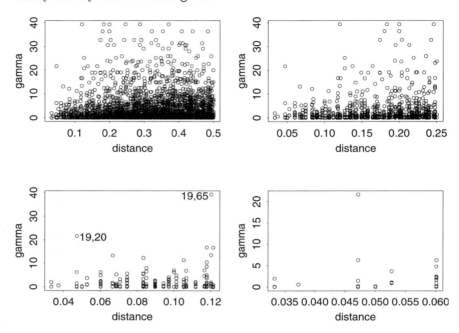

FIGURE 4.7. Variogram clouds for scallops data at various maximum distances.

Location pairs of interest can be interactively identified by using **identify**. The dissimilar (large gamma for the given distance) locations (19,20) and (19,65) identified in figure 4.7 are obtained as follows:

```
> identify(scallops.vcloud5)
```

Once invoked, **identify** allows you to continuously select location pairs by clicking the left mouse button at each point of interest. To exit, click the middle mouse button (right mouse button in Windows) while in the graphics screen. By default, the location pairs are identified on the plot and listed in the S-PLUS command window. The list of points can be saved in an S-PLUS object as follows:

```
> scall.prs <- identify(scallops.vcloud5)
```

Note: In this example, the **identify** function was called before plotting the **scallops.vcloud6** object.

To view a summary while still being able to look for potential outliers, create a boxplot from the variogram cloud as shown in figure 4.8:

```
> boxplot(scallops.vcloud1)
```

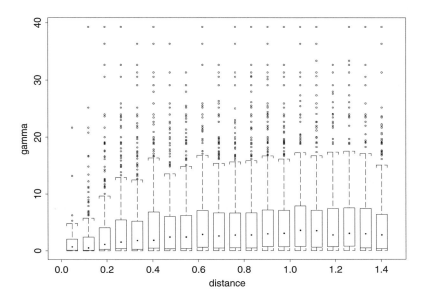

FIGURE 4.8. Boxplots of the squared-difference cloud for scallops data.

By default the data is separated into 20 bins. Many of the points fall outside the top whiskers, indicating potential outliers. The outlying points may be due to the skewed distribution of the variogram cloud as opposed to atypical observations. To check this, create boxplots based on the square-root-differences cloud as follows:

```
> scallops.vcloud2 <- variogram.cloud(log(tcatch+1)
+       ~ loc(lat,long), data=scallops,
+       fun=function(zi,zj) sqrt(abs(zi-zj))/2)
> boxplot(scallops.vcloud2, mean=T, pch.mean="o")
```

The optional **fun** argument is used to calculate the square-root-differences. The **mean** and **pch.mean** arguments are used to include means, in addition to medians, on the boxplots. The boxplots of the square-root-differences cloud are shown in figure 4.9. Only three points are left as potential outliers, all of which occur at distances less than 0.2.

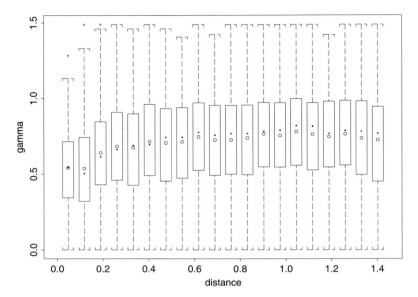

FIGURE 4.9. Boxplots of the square-root-differences cloud for the scallops data.

4.1.3 Detecting and Removing Trends

The existence of the variogram is based on the assumption of *intrinsic stationarity* of the random function, which is defined through first differences as follows:

$$E(Z(i+h) - Z(i)) = 0, \quad \text{for all } i, i+h \in D,$$

and

$$var(Z(i+h) - Z(i)) = 2\gamma(h).$$

Essentially, intrinsic stationarity implies a process with a constant mean and with a variance defined only through the magnitude of h.

Models of geostatistical data, however, are often composed of both large-scale trend or drift, and small-scale random variation. In this case, the random field, $Z(x)$, does not have constant mean, and a variogram based on $Z(x)$ will not meet the necessary assumption.

Detection of possible trends often results from exploratory data analysis. Some graphical methods for elucidating possible trends are shown in sections 3.2.1 and 3.2.2. In section 4.1.1 of this chapter, the use of directional variograms is shown as a way to detect possible trend. Regardless of how a trend is detected, the goal is to appropriately model and detrend the data, and then estimate the variogram for the underlying random process.

Median Polishing

Median polishing is a resistant method for detrending gridded data and is based on an additive decomposition, where

$$\text{data} = \textit{grand} + \textit{row} + \textit{column} + \textit{residual}.$$

Since median polishing assumes an additive trend, the method is not appropriate for trend models with interaction effects between rows and columns. The algorithm iterates successive sweeps of medians out of rows, then out of columns, accumulating them in *row*, *column*, and *grand* effects. The *residuals* are what is left after the algorithm converges.

Median polishing requires data aligned in rows and columns, and thus is naturally suited for use with gridded data; median polishing can be used on non-gridded data only when coerced to a grid. Use the function `twoway` to perform a median polish on the coal ash data as follows:

```
> coal.mp <- twoway(coal~x+y,data=coal.ash)
```

By default, `twoway` will sweep the rows and columns of the coal ash data using medians. Other variations of row and column sweeping, including mean polishing, can be done by setting the optional `trim` argument.

Note: We are using the new generic `twoway` function included with S+SPATIALSTATS. The `twoway.formula` function gets called here.

The call to `twoway` returns the grand median and vectors of row effects, column effects, and residuals. The residuals, in conjunction with the original data, can be used to look for trend as follows:

1. Subtract the median polish residuals from the original data to capture the signal.

   ```
   > coal.signal <- coal.ash$coal-coal.mp$residuals
   ```

2. Convert the vectors for the original data and signal to matrices for use with the `image` function.

   ```
   > coal.mat <- tapply(coal.ash$coal,list(
   +       factor(coal.ash$x),factor(coal.ash$y)),
   +       function(x)x)
   > coalsig.mat <- tapply(coal.signal,list(
   +       factor(coal.ash$x),factor(coal.ash$y)),
   +       function(x)x)
   ```

3. Set up the plotting device and limits for the *z* values.

```
> motif()
> par(mfrow=c(1,2))
> # force a square plotting region
> par(pty="s")
> zmin <- min(coal.mat[!is.na(coal.mat)],
+      coalsig.mat[!is.na(coalsig.mat)])
> zmax <- max(coal.mat[!is.na(coal.mat)],
+      coalsig.mat[!is.na(coalsig.mat)])
```

4. Produce a grayscale plot of the original data and the signal using a common color scale.

```
> image(coal.mat, zlim=c(zmin,zmax))
> image(coalsig.mat, zlim=c(zmin,zmax))
```

The resulting plots are shown in figure 4.10. An apparent trend in the east-west direction can be seen in the **image** plot of the signal (right-hand plot).

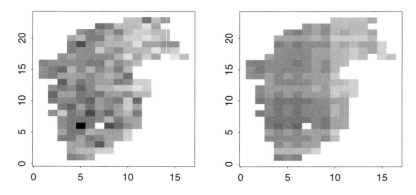

FIGURE 4.10. Results of median polishing of the coal ash data, showing image plots of the original data (left) and the apparent E-W trend in the signal (right).

The effect of the median polish trend removal can be seen by looking at a variogram of the residuals in the east-west direction and comparing it with the corresponding variogram from the original data as follows:

1. Create east-west variograms based on the median polish residuals and the original data, using a common y-axis scale.

```
> coal.var5 <- variogram(coal.mp$residuals
+      ~ loc(x,y),data=coal.ash,
+      azimuth=90, tol.azimuth=11.25)
```

```
> coal.var6 <- variogram(coal~loc(x,y),data=coal.ash,
+         azimuth=90, tol.azimuth=11.25)
```

2. Plot the east-west variograms.

```
> par(mfrow=c(2,1))
> ymax <- max(coal.var5$gamma,coal.var6$gamma)
> plot(coal.var5,ylim=c(0,ymax))
> plot(coal.var6,ylim=c(0,ymax))
```

The east-west variograms based on the median polish residuals and the original data are shown in figure 4.11. Based on the variograms, most of the correlation apparent in the east-west direction seems to be due to the trend.

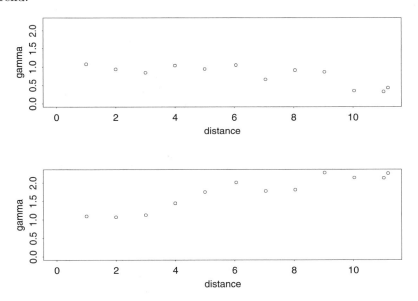

FIGURE 4.11. East-west variogram of the coal ash data calculated from median polish residuals (top) and original data (bottom).

Other Methods

In section 3.2.2, a generalized additive model was fitted to the scallop catch values using smooth functions of latitude and longitude as predictors. Plots of the resulting smooth functions were used to visualize possible trends. Given the observation of trends, a spatial loess model was fitted to the data using the latitude and longitude as predictors. Residuals from the loess model can be evaluated in a variogram analysis.

Ordinary and generalized linear models are additional methods that may be used to model and remove trend from data.

4.1.4 Anisotropy

Anisotropy is present when the spatial autocorrelation of a process changes with direction; the underlying physical process evolves differently in space. Unlike a variogram from an *isotropic* process, the variogram from an anisotropic process is not purely a function of the distance h, but is a function of both the magnitude and direction of $\mathbf{h}$.

There are two types of anisotropy: *geometric anisotropy* occurs when the range of the variogram changes in different directions, while the sill remains constant; *zonal anisotropy* exists when the sill of the variogram changes with direction.

Identifying and correcting for anisotropy are important because the theoretical variograms used for kriging are based on isotropic models. Geometric anisotropy is generally corrected by a linear transformation of the spatial locations to an equivalent isotropic model. Zonal anisotropy may be corrected by appropriately modeling and detrending the data, or by choosing a *nested* variogram model. One component of a nested model would be an isotropic model fitted to the direction of the zonal anisotropy; the other component might be a geometrically anisotropic variogram function.

Identifying Anisotropy

Directional variograms created with **variogram** can be used to detect anisotropy. Continuing with the rotated scallops data first introduced in section 3.2.2, figure 4.12 shows some directional variograms created as follows:

```
> scallops.dvar1 <- variogram(lgcatch ~ loc(newx,newy),
+       data=scall.rot, azimuth=c(0,45,90,135),
+       tol.azimuth=11.25)
> plot(scallops.dvar1)
```

While the 45 and 135 degree directions yield similar variograms, there are apparent anisotropies in the 0 and 90 degree directions. In the rotated space, the 0 and 90 degree directions correspond roughly to a perpendicular and a parallel coastline orientation, respectively.

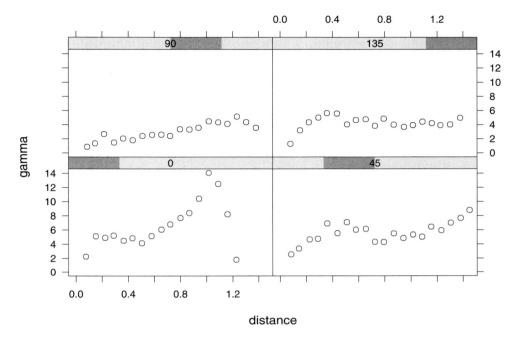

FIGURE 4.12. Directional variograms for rotated scallops data showing anisotropies in the 0 and 90 degree directions.

Disregarding the last four points (due to the low number of pairs) of the 0 degree directional variograms, yields an increasing variogram with no bounds. This could be caused by the apparent trend that was identified through exploratory data analysis using generalized additive models (section 3.2.2).

The 90 degree variogram seems to represent a case of geometric anisotropy. It has a sill of ≈ 5, which looks similar to the sills of the 45 and 135 degree variograms; while its apparent range (≈ 1) is almost twice that of the 45 and 135 degree variograms. Evidently, the autocorrelation parallel to the coastline has a greater range than that for the other directions. This may be expected, given similar environmental conditions parallel to the coastline as opposed to perpendicular to it.

Correcting for Anisotropy

If the anisotropy in the 0 degree variogram in figure 4.12 is the result of trend, then the directional variogram based on detrended data should indicate this. Figure 4.13 shows the same directional variograms, created

using the residuals from a spatial loess model of the rotated scallops data
(see section 3.2.2 for a description of the loess model).

```
> scall.res <- scall.rot$lgcatch - predict(loess.scp)
> scallops.dvar2 <- variogram(scall.res
+        ~ loc(scall.rot$newx, scall.rot$newy),
+        azimuth=c(0,45,90,135), tol.azimuth=11.25,
+        method="robust")
> plot(scallops.dvar2)
```

The 0 degree variogram is no longer increasing and appears to have a range
and sill similar to the 45 degree variogram. The 90 degree variogram still
has a range considerably greater than the variograms in other directions.
The 135 degree variogram may be a pure nugget effect, showing little or
no spatial correlation. The sills of all the variograms are approximately
the same and are reduced from the original directional variograms. This is
consistent with the removal of trend.

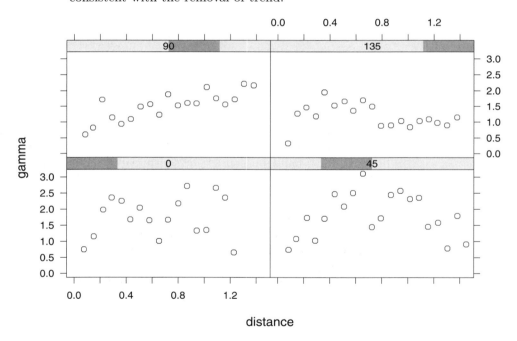

FIGURE 4.13. Directional variograms for residuals from a loess model on the
rotated scallops data.

To better define the geometric anisotropy, a description of how the range
changes in various directions is needed. This can be evaluated by looking

at contours of gamma for various distances and directions. One way to do this is as follows:

1. Write a panel function that calculates the distance value for a specified value of gamma. Use a loess smooth to approximate the variogram function and then the `approx` function to determine the distance corresponding to a given value of gamma on the loess curve.

```
> panel.gamma0 <-
+ function(x, y, gamma0 = gamma0, span = 2/3, ...)
+ {
+     lofit <- loess.smooth(x, y, span = span)
+     panel.xyplot(x, y, ...)
+     panel.xyplot(lofit$x, lofit$y, type = "l")
+     dist0 <- approx(lofit$y, lofit$x,
+         xout = gamma0)$y
+     segments(0, gamma0, dist0, gamma0)
+     segments(dist0, 0, dist0, gamma0)
+     parusr <- par()$usr
+     text(parusr[2] - 0.05 * diff(parusr[1:2]),
+         parusr[3] + 0.05 * diff(parusr[3:4]),
+         paste("d0=", format(round(dist0, 4)),
+         sep = ""), adj = 1)
+ }
```

For convenience, this panel function is included with S+SPATIAL-STATS; however, a help file is not provided.

2. Subset `scallops.dvar2` to include only points for distances less than 0.9 in the 0 and 45 degree directions, due to the low numbers of pairs.

```
> scallops.aniso <- scallops.dvar2[
+     (scallops.dvar2$distance < 0.9 &
+     scallops.dvar2$azimuth == 0) |
+     (scallops.dvar2$distance < 0.9 &
+     scallops.dvar2$azimuth == 45) |
+     scallops.dvar2$azimuth == 90 |
+     scallops.dvar2$azimuth == 135,]
```

3. Plot the variograms with interpolated distances for gamma = 1.4, using `xyplot` with the panel function `panel.gamma0`.

```
> xyplot(gamma ~ distance | azimuth,
+     data=scallops.aniso, panel=panel.gamma0,
+     gamma0=1.4)
```

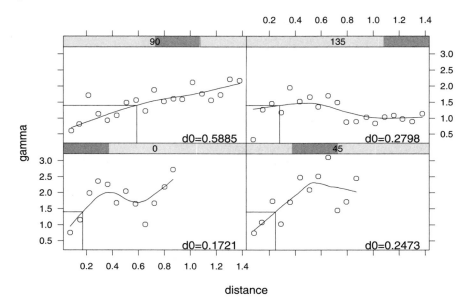

FIGURE 4.14. Directional variograms for scallops loess residuals with interpolated distances for $\gamma(1.4)$.

Figure 4.14 shows the interpolated distances for $\gamma(1.4)$.

The interpolated distances, h, are similar for the 45 and 135 degree directions. The most prominent difference is in the 90 degree direction.

Note: The independence assumption for the loess smoother is not met for the variogram values. It is only used to get a first approximation, which seems sufficient to gauge the anisotropy.

The interpolated distances can be used to construct a rose diagram (Isaaks and Srivastava, 1989), providing another technique for visualizing and defining the anisotropy. Rose diagrams are equivalent to showing the distance in several directions along a given gamma contour. Create the rose diagram shown in figure 4.15 as follows:

1. Scale the interpolated distances relative to the largest value of h. In this case, the scaled distances are .292, .420, 1.000, and .475 for 0 through 135 degrees, respectively:

```
> dscale <- .5885
> d0 <- .1721/dscale
> d45 <- .2473/dscale
> d90 <- 1
> d135 <- .2798/dscale
```

2. Determine the points on the unit xy surface that correspond to the scaled distances in each direction. For the 0 and 90 degree directions, the points are (0,.292) and (1,0). For the 45 degree direction, the point can be calculated from the equation for a circle, $x^2 + y^2 = r^2$, where r is the radius, or the distance in this case. For an azimuth of 45 degrees, $x = y$, so the calculation simplifies to $x^2 + x^2 = r^2$. The point for the 45 degree direction is (.297,.297). The point for the 135 degree direction is (.336,.336).

3. Use the functions `plot` and `segments` to generate the rose diagram:

```
> plot(0,0,type='n', axes=F, xlim=c(-1,1),
+      ylim=c(-1,1),
+      xlab="along scall.rot$newx",
+      ylab="along scall.rot$newy")
> segments(0,0,1,0)
> segments(0,0,0,.292)
> segments(0,0,.297,.297)
> segments(0,0,.336,-.336)
> segments(0,0,-1,0)
> segments(0,0,0,-.292)
> segments(0,0,-.297,-.297)
> segments(0,0,-.336,.336)
```

Rose diagrams are symmetric (since variograms are symmetric); the segments are extended in corresponding 180 degree directions to complete the diagram.

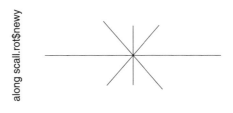

FIGURE 4.15. Rose diagram for scallops loess model residuals based on interpolated distances for $\gamma(1.4)$.

The shape of the rose diagram is approximately elliptical with the major axis in the 90 degree direction. This further defines the form of the geometric anisotropy, which can be corrected by a linear transformation.

S+SPATIALSTATS provides the function `anisotropy.plot` to perform linear transformations of the data and to generate the resulting isotropic variograms. Correct for the geometric anisotropy in the scallops data as follows:

```
> anisotropy.plot(scall.res ~ loc(scall.rot$newx,
+     scall.rot$newy), angle=90,
+     ratio=c(2.25,2.5,2.75,3.0,3.25,3.5),
+     method="robust", layout=c(2,3))
```

The optional argument `angle` is the clockwise angle of rotation, in degrees, by which the y axis must be rotated to become parallel to the major axis of the ellipse. The optional argument `ratio` is the ratio of the length of the major to minor axes of the ellipse. For the scallops data, the axes must be rotated a full 90 degrees, so we set `angle` to 90. Based on a gamma contour of 1.4, the ratio of the major to minor ellipse axes is approximately 3. Thus, we set `ratio` to a range of values around 3 for comparison. Figure 4.16 shows the variograms resulting from the linear transformations.

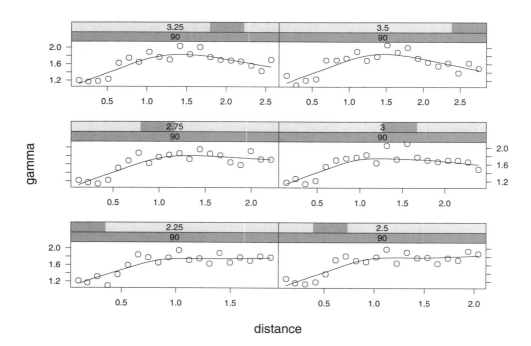

FIGURE 4.16. Scallops variograms corrected for trend and geometric anisotropy.

The resulting variograms, corrected for geometric ansiotropy and trend, are similar. The specific choice as to which variogram to use for model fitting is up to the researcher.

4.2   Modeling the Empirical Variogram

In section 4.3 we show how the variogram is incorporated into kriging equations for use in making predictions and calculating kriging prediction variances. In order to ensure that the variance of predicted values is positive, the empirical variogram must be replaced with a theoretical variogram function.

4.2.1   Theoretical Variogram Models

S+SPATIALSTATS provides functions for the common theoretical variogram models. The *exponential, spherical,* and *gaussian* models are bounded variogram functions. The *linear* and *power* models increase without bounds. Plot examples of the various theoretical models shown in figure 4.17 as follows:

```
> par(mfrow=c(3,2))
> vdist <- 1:15
> vrange <- 7
> plot(vdist,exp.vgram(distance=vdist, range=vrange),
+      type="l")
> plot(vdist,spher.vgram(distance=vdist, range=vrange),
+      type="l")
> plot(vdist,gauss.vgram(distance=vdist, range=vrange),
+      type="l")
> plot(vdist,linear.vgram(distance=vdist, slope=.3),
+      type="l")
> plot(vdist,power.vgram(distance=vdist, slope=.3,
+      range=.5), type="l")
```

All of the theoretical variogram functions require specification of a `distance` vector. The exponential, spherical and gaussian models also require a `range` value. The linear and power models require a `slope` value. The `nugget` argument is optional for all models and defaults to 0. In addition, there is an optional `sill` argument for the bounded functions (default = 1) and a `range` argument required for the power function.

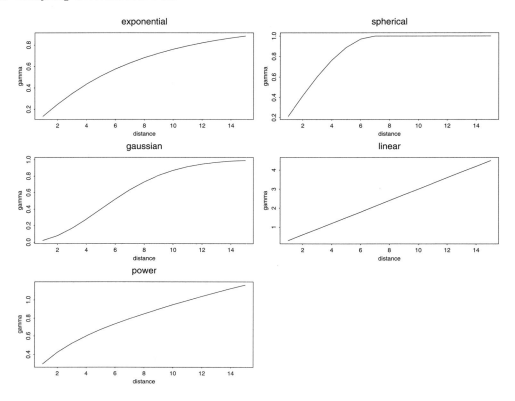

FIGURE 4.17. Theoretical variogram models in S+SPATIALSTATS.

For `exp.vgram` the range is defined as approximately one-third the *apparent* range, where the apparent range is the observed distance h above which data no longer appear to be correlated. For the spherical and gaussian models, the `range` argument is the apparent range. For the power model, the `range` argument is the exponent of the distance. The `sill` argument represents the *absolute* sill; the absolute sill is estimated as the sill of the empirical variogram minus any nugget effect.

4.2.2   Fitting a Theoretical Variogram Model

Fitting a theoretical variogram model to an empirical variogram is often done by eye. Even if a numerical fitting routine is to be used, looking at an initial fit by eye can be useful. The initial model type is chosen based on the shape of the empirical variogram and the researcher's belief as to the nature of the operating processes. Initial values for the range, sill, and nugget effect parameters are also chosen from the empirical variogram. The fit of the theoretical model can be viewed and interactively updated using the S+SPATIALSTATS function `model.variogram`. Within a call to `model.variogram`, parameter values can be modified, and the results overlaid onto the empirical variogram until a satisfactory fit is obtained.

Fit a theoretical variogram model to the scallops data variogram that was corrected for geometric anisotropy (section 4.1.4) as follows:

1. Calculate the empirical variogram based on the appropriate correction (transformation) for geometric anisotropy. Based on the plots in figure 4.16, a correction using an angle of 90 degrees with a ratio of 2.25 yields a reasonable variogram. Create the transformed variogram as follows:

   ```
   > scallops.finalvar <- variogram(scall.res
   +       ~loc(scall.rot$newx, scall.rot$newy,
   +       angle=90, ratio=2.25), method="robust")
   ```

2. Choose a theoretical model and initial parameter values. Based on the general shape of the empirical variogram (figure 4.16, bottom left), we will start with a spherical model having range = .8, sill = 1.75, and nugget effect = .5.

3. Call `model.variogram` with the initial model choice.

   ```
   > model.variogram(scallops.finalvar,fun=spher.vgram,
   +             range=.8, sill=1.75-.5, nugget=.5)
   Select a number to change a parameter (or 0 to exit):
    Current objective = 0.4814
   1: range - current value: 0.8
   2: sill - current value: 1.25
   3: nugget - current value: 0.5
   Selection:
   ```

The `model.variogram` function displays the initial theoretical and empirical variograms (figure 4.18, top left) and an interactive menu that allows changes to be made in the parameter values for the current model type. Choose options 1, 2, or 3 to modify the range, sill, or nugget parameters for the spherical variogram. After each change, the updated theoretical model is automatically plotted. The `model.variogram` function also includes a measure of model fit through the optional argument `objective.fun`. By default, the objective function is the residual sum of squares between the theoretical model and the empirical variogram. For each iteration of parameter changes, a new objective value is calculated and displayed on both the command screen and the plot. The user may implement a custom objective function. Enter 0 at the `Selection:` prompt to exit `model.variogram`.

4. Adjust the fit, if necessary, by modifying parameter values. For example, it looks as though we may need to increase the nugget effect (and appropriately adjust the sill) and decrease the range. Do this as follows:

```
Selection: 3
New nugget : .7
Select a number to change a parameter (or 0 to exit):
 Current objective = 0.9705
1: range - current value: 0.8
2: sill - current value: 1.15
3: nugget - current value: 0.7
Selection: 2
New sill : 1.05
Select a number to change a parameter (or 0 to exit):
 Current objective = 0.3352
1: range - current value: 0.8
2: sill - current value: 1.05
3: nugget - current value: 0.7
Selection: 1
New range : .75
Select a number to change a parameter (or 0 to exit):
 Current objective = 0.3522
1: range - current value: 0.75
2: sill - current value: 1.05
3: nugget - current value: 0.7
Selection: 0
```

The resulting fitted models are shown in figure 4.18. Using the interactive menu, the nugget was first increased to 0.7 (top right); to compensate for the increased nugget, the sill was decreased to 1.05 (bottom left); the range was decreased to 0.75 (bottom right). The residual sum of squares decreased with each iteration of a fitted model, until the range was decreased. The final model of choice is a spherical variogram with range = 0.8, sill = 1.75, and nugget effect = 0.7.

The function `model.variogram` can also be used to interactively fit covariograms and correlograms.

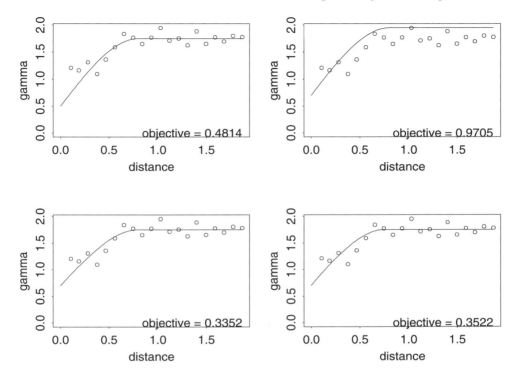

FIGURE 4.18. Scallops empirical variogram with several iterations resulting in a fitted spherical model.

Fitting Using Nonlinear Least Squares

Variogram models can be fit by optimization techniques, typically some form of nonlinear least squares. The usual statistical assumptions for nonlinear regression models are not valid for fitting variograms since the variogram values at different lags are not independent. Cressie (1985) describes weighted least squares and generalized least squares approaches for variogram fitting that try to account for the special structure of the variogram values. Zimmerman and Zimmerman (1991) compare several estimation methods and conclude that ordinary nonlinear least squares or some form of weighted nonlinear least squares is usually as good as many of the more complicated and computationally intensive methods.

The use of nonlinear least squares variogram fitting is demonstrated on the coal ash data. Earlier (section 3.2.1) it was determined that there is a trend in the east-west direction of this data. For illustration purposes, we will fit a spherical variogram model to only the north-south data as follows:

1. Compute and plot the north-south variogram.

   ```
   > coal.varns <- variogram(coal ~ loc(x, y),
   +        data = coal.ash, azimuth = 0,
   +        tol.azimuth = .01, lag = 1)
   > plot(coal.varns)
   ```

   The resulting variogram is shown in figure 4.19.

2. Write a function defining the residuals for which the sums of squares will be minimized, using the spherical variogram function spher.vgram.

   ```
   > spher.fun <- function(gamma, distance, range,
   +        sill, nugget)
   +        gamma - spher.vgram(distance, range = range,
   +        sill = sill, nugget = nugget)
   ```

3. Choose starting values. From the variogram plot of coal.varns, we choose range = 4.0, sill = 0.2, and nugget effect = 0.8.

4. Call nls using the function spher.fun defined above.

   ```
   > coal.nl1 <- nls( ~ spher.fun(gamma, distance,
   +        range, sill, nugget), data = coal.varns,
   +        start = list(range = 4, sill = 0.2, nugget=.8))
   > coef(coal.nl1)
        range      sill    nugget
     3.443335 0.240059 1.093492
   ```

5. Add the fitted model to the empirical variogram (figure 4.19).

   ```
   > lines(coal.varns$dist, spher.vgram(coal.varns$dist,
   +        range=3.443335,sill=.240059,nugget=1.093492))
   ```

Cressie (1985) suggests minimizing the weighted sum of squares :

$$\sum_{j=1}^{K} |N(h(j))| \left\{ \frac{\gamma(h(j))}{\gamma(h(j); \boldsymbol{\theta})} - 1 \right\}^2 ,$$

where $|N(h(j))|$ is the number of distinct pairs in lag j, K is the number of lags in the empirical variogram, $\gamma(h(j))$ is the value of the empirical variogram at lag j, and $\gamma(h(j); \boldsymbol{\theta})$ is the known variogram model with unknown parameters $\boldsymbol{\theta}$.

This weighting function can be easily incorporated into nls by defining the residual function to include the weights:

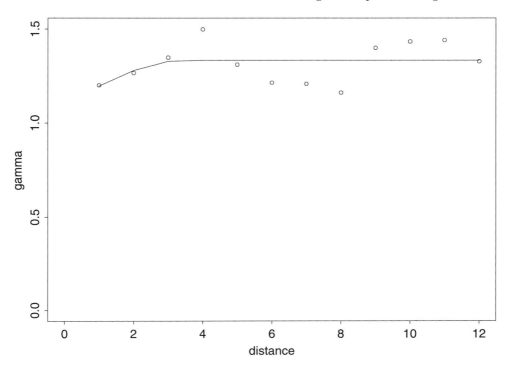

FIGURE 4.19. Coal ash empirical variogram (open circles) with a spherical fitted model (line) based on estimates using nonlinear least squares.

```
> spher.wfun <-
+ function(gamma, distance, np, range, sill, nugget)
+ {
+     gammahat <- spher.vgram(distance, range = range,
+     sill = sill, nugget = nugget)
+          sqrt(np) * (gamma/gammahat - 1)
+ }
```

Calculate new variogram parameter estimates for the coal ash data, incorporating the weights, as follows:

```
> coal.nl2 <- nls( ~ spher.wfun(gamma, distance,
+       np, range, sill, nugget), data = coal.varns,
+       start = list(range = 4, sill = 0.2, nugget=.8))
> coef(coal.nl2)
    range       sill     nugget
 3.658897 0.2427316 1.099064
```

In this case, the coefficients from the weighted nonlinear least squares are not much different than the unweighted coefficients.

Other functions for nonlinear fitting in S-PLUS include `ms` and `nlminb`. See the individual help files for details on these functions. For additional information on fitting nonlinear models in S-PLUS see Bates and Chambers (1992) or Venables and Ripley (1994).

4.3   Kriging

Kriging is a linear interpolation method that allows predictions of unknown values of a random function from observations at known locations. Kriging incorporates a model of the covariance of the random function when calculating predictions of the unknown values. S+SPATIALSTATS provides functions to perform two types of kriging: *ordinary* and *universal*. Ordinary kriging uses a random function model of spatial correlation to calculate a weighted linear combination of available samples, for prediction of a nearby unsampled location. Weights are chosen to ensure that the average error for the model is zero and that the modeled error variance is minimized (Isaaks and Srivastava, 1989). Universal kriging is an adaptation of ordinary kriging that accomodates trend. Universal kriging can be used to both produce local estimates in the presence of trend, and to estimate the underlying trend itself. Universal kriging with a constant mean is equivalent to ordinary kriging.

4.3.1   Ordinary Kriging

Ordinary kriging in 2-dimensions is performed in S+SPATIALSTATS by using the `krige` and `predict.krige` functions: `krige` uses the kriging response variable, spatial locations, and a theoretical covariance function to set up kriging matrices for the predictions; `predict.krige` uses the output from `krige` to compute kriging predictions and standard errors for unsampled locations specified by the user. The theoretical covariance function is based on a theoretical variogram, and is defined for the exponential, spherical and gaussian models. The linear and power variogram models do not have corresponding covariance models since they are unbounded. A linear covariance model can be calculated, however, by subtracting the values of a linear variogram from a large number (the value of gamma corresponding to the greatest distance over which predictions will be made).

Perform ordinary kriging of the scallops data as follows:

```
> scallops.krige <- krige(scall.res~loc(newx,newy,
+     90,2.25), data=scall.rot, covfun=spher.cov,
+     range=.8, sill=(1.75-.7), nugget=.7)
```

The kriging variable is `scall.res`, the residuals from a spatial loess model. The spatial locations are the rotated coordinates `newx` and `newy`. The spatial correlation is modeled as spherical covariance based on the spherical variogram model fitted to the empirical variogram (figure 4.18, bottom left). As with `spher.vgram`, the `sill` argument for `spher.cov` is specified as the sill minus the nugget effect.

The `krige` function returns an object of class `"krige"` that includes a summary of the call and the calculated coefficients:

```
> scallops.krige
Call:
krige(formula = scall.res ~ loc(newx, newy, 90, 2.25),
    data = scall.rot, covfun = spher.cov, range = 0.8,
    sill = (1.75 - 0.7), nugget = 0.7)

Coefficients:
   constant
 -0.1873672

Number of observations: 148
```

Kriging predictions at a set of unsampled spatial locations can now be made using `predict.krige`. There are two ways to define unsampled locations for the prediction. One way involves the `newdata` argument; `newdata` is a data frame or list containing the spatial locations for the predictions. Alternatively, a grid of points can be generated using the `grid` argument; `grid` is a list of two vectors (one for each axis), specifying the minimum, maximum, and number of points. For either `newdata` or `grid`, the names must match the names of the locations used in the call to `krige`. By default, a set of prediction locations are generated on a 30 × 30 grid defined by the minimum and maximum spatial coordinates of the sampled locations. Calculate kriging predictions for the scallops residuals using the default locations as follows:

```
> scallops.pkrige1 <- predict(scallops.krige)
```

The generic `predict` function calls the method `predict.krige` for an object of class `"krige"`. The call to `predict` returns a data frame with four columns corresponding to the x, y locations of the predictions, the predictions, and the standard errors of the predictions. Plot the sampled and prediction locations as follows:

```
> plot(scall.rot$newx, scall.rot$newy,
+      xlab="newx", ylab="newy", pch=16)
> points(scallops.pkrige1$newx, scallops.pkrige1$newy,
+      pch="+")
```

Figure 4.20 shows the original sampled locations and the default prediction locations returned from `predict`.

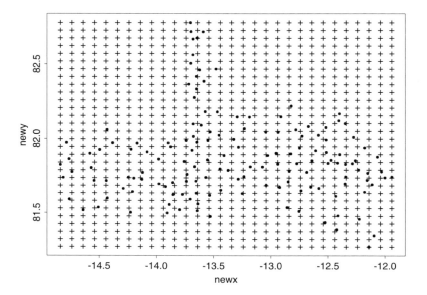

FIGURE 4.20. Original scallop sampling locations (closed circles) and default prediction locations (+'s).

These prediction locations may not be satisfactory for the scallops data. There are too many locations close to shore, far from sampled locations; some of the default locations are on land. In order to use the `grid` argument, minimum and maximum locations corresponding to the main rectangular sampling area should be used. Alternatively, S+SPATIALSTATS and S-PLUS have several functions that can be used to generate prediction locations within polygonal boundaries. The function `chull` can be used to calculate the convex hull of a sampling region; `locator` can be used to interactively construct a user-defined polygon around, or within a sampling region. Create the polygonal boundaries shown in figure 4.21 as follows:

```
> par(mfrow=c(1,2))
> par(pty="s")
>
```

```
> plot(scall.rot$newx, scall.rot$newy, pch=16)
> scallops.chull <- chull(scall.rot$newx, scall.rot$newy)

> polygon(scall.rot$newx[scallops.chull],
+        scall.rot$newy[scallops.chull], density=0)
>
> plot(scall.rot$newx, scall.rot$newy, pch=16)
> scallops.poly <- locator(type = "l")
```

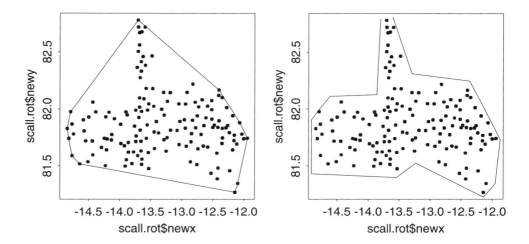

FIGURE 4.21. Convex hull (left) and user-defined polygon (right) based on scallops data sampling locations.

The call to `chull` returns the indices of the location vectors which correspond to the vertices of the convex hull. The `polygon` function adds the convex hull to the current plot. By default, `polygon` shades in the convex hull; setting the optional `density` argument to 0 causes only the outline to be drawn. The `locator` function, with the optional `type` argument set to `"l"`, interactively draws a polygon as user selected points are identified. Select boundaries for the polygon using the *left* mouse button. Exit `locator` from within the graphics window by using the *middle* mouse button in UNIX or the *right* mouse button in Windows. There is no need to close a user-defined polygon.

Prediction locations can be defined within the bounds of the convex hull or user-defined polygon using `poly.grid`. Create the prediction locations shown in figure 4.22 as follows:

```
> par(mfrow=c(1,2))
> par(pty="s")
```

```
>
> plot(scall.rot$newx, scall.rot$newy, pch=16)
> predict.loc1 <- poly.grid(cbind(
+       scall.rot$newx[scallops.chull],
+       scall.rot$newy[scallops.chull]), nx=20, ny=20)
> points(predict.loc1,pch="+")
>
> plot(scall.rot$newx, scall.rot$newy, pch=16)
> predict.loc2 <- (poly.grid(scallops.poly,
+       size=c(0.08,0.05)))
> points(predict.loc2, pch="+")
```

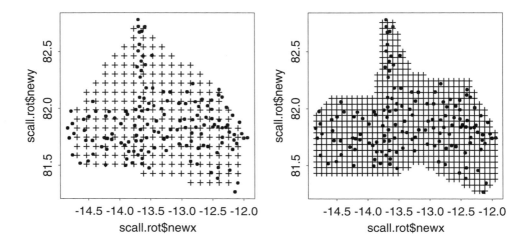

FIGURE 4.22. Prediction locations for the convex hull (+'s, left) and user-defined polygon (+'s, right).

In the left plot (figure 4.22) the prediction locations were specified by setting the **nx** and **ny** arguments of **poly.grid** resulting in a $nx \times ny$ grid. The locations on the right plot were created using the optional **size** argument to create a grid with spacing of 0.08 in the x direction and 0.05 in the y direction.

Perform ordinary kriging of the scallops residuals, using the prediction locations within the user-defined polygon in figure 4.22, as follows:

1. Convert **predict.loc2**, the output from a call to **poly.grid**, to a data frame. Change the names of the prediction locations to match those used in the call to **krige**.

Note: A data frame is not required for the function `predict.krige`, but a data frame is required for `predict.loess` which will be called later.

```
> predict.loc2 <- as.data.frame(predict.loc2)[c(1,2)]
> names(predict.loc2) <- c("newx","newy")
```

2. Calculate predictions.

```
> scallops.pkrige2 <- predict(scallops.krige,
+         newdata=predict.loc2)
```

3. Look at a summary of the predictions.

```
> summary(scallops.pkrige2)
      newx                 newy                 fit
 Min.   :-14.83     Min.   :81.28      Min.   :-3.07000
 1st Qu.:-13.92     1st Qu.:81.58      1st Qu.:-0.33770
 Median :-13.43     Median :81.84      Median :-0.09671
 Mean   :-13.36     Mean   :81.85      Mean   :-0.22850
 3rd Qu.:-12.69     3rd Qu.:82.04      3rd Qu.: 0.19210
 Max.   :-11.95     Max.   :82.75      Max.   : 0.96700

      se.fit
 Min.   :0.9445
 1st Qu.:1.0030
 Median :1.0320
 Mean   :1.0490
 3rd Qu.:1.0800
 Max.   :1.2540
```

In this case, `newx` and `newy` are the prediction locations, `fit` is the kriging predictions for the residuals, and `se.fit` is the corresponding kriging prediction standard errors. To generate a predicted surface for the scallops catch data, add predictions from the spatial loess model to the kriged residuals as follows:

```
> scall.lo <- predict(loess.scp, predict.loc2)
> scall.pred <- scallops.pkrige2$fit + scall.lo
```

Create contour maps for the predicted surface and relative standard errors as follows:

1. Map the scallop predictions and kriging prediction standard errors to matrices bounded by the locations in `predict.loc2`.

```
> xmat <- sort(unique(predict.loc2$newx))
> ymat <- sort(unique(predict.loc2$newy))
> scall.predmat <- matrix(NA, length(xmat),
+     length(ymat))
> scall.predmat[cbind(match(predict.loc2$newx,
+     xmat), match(predict.loc2$newy, ymat))] <-
+     scall.pred - 1
> scall.semat <- matrix(NA, length(xmat),
+     length(ymat))
> scall.semat[cbind(match(predict.loc2$newx,
+     xmat), match(predict.loc2$newy, ymat))] <-
+     scallops.pkrige2$se.fit
```

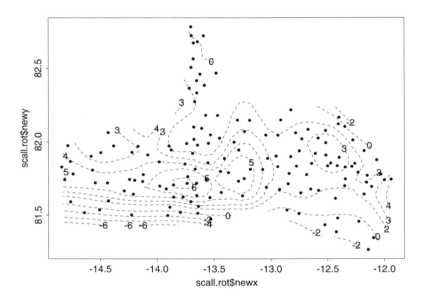

FIGURE 4.23. Contour plot of kriging predictions for logged scallops total catch.

2. Plot the original sampling locations. Overlay contours based on kriging predictions and the spatial loess model. Plot contours for the relative standard errors.

```
> plot(scall.rot$newx, scall.rot$newy)
> contour(xmat,ymat,scall.predmat,
+     levels=c(-6,-4,-2,0,2,3,4,5,6),
+     add=T,lty=3)
> plot(scall.rot$newx, scall.rot$newy)
> contour(xmat,ymat,scall.semat)
```

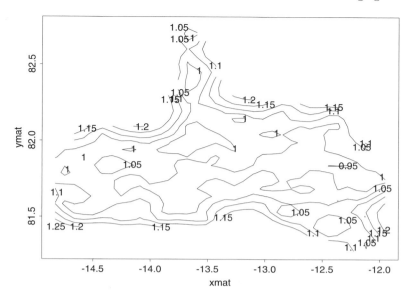

FIGURE 4.24. Contour plot of relative kriging prediction standard errors for logged scallops total catch.

The plots are displayed in figures 4.23 and 4.24.

4.3.2   Universal Kriging

Universal kriging is performed in S+SPATIALSTATS using the same functions as for ordinary kriging; namely, `krige` and `predict.krige`. The difference is in the formula specification for `krige`. Since universal kriging simultaneously fits a trend model to the data, the formula must contain a specification for a polynomial trend surface.

Universal kriging requires knowledge of both a trend model and a variogram or covariance model for the data. The variogram cannot be correctly estimated if there is a trend and the trend cannot be estimated correctly by standard methods if there is spatial correlation. Often an iterative approach is used; initial covariance and trend models are selected and universal kriging is run. The residuals from the trend surface (`resid(obj)`, where `obj` is an object returned from `krige`) can be examined to see if the trend model is correctly specified. The variogram of these residuals can also be used to refine the covariance model. Universal kriging is then run again with the refinements to the model.

In some situations, additional information may allow estimation of the variogram from subsets of the data. With the appropriate assumptions, this model can then be used in the universal kriging of the entire data set. This is illustrated in the example below.

The use of universal kriging is demonstrated using the coal ash data first introduced in section 3.2.1. Exploratory data analysis showed an apparent trend in the east-west direction. Earlier in this chapter (section 4.1.3) the trend was removed using median polishing. Resulting directional variograms showed essentially no remaining spatial correlation in the east-west direction. As such, the spatial correlation is modeled using a spherical variogram fitted to the empirical variogram in the north-south direction. The parameters were estimated using nonlinear least squares in section 4.2.2. The trend model is specified based on the observed trend in the east-west direction (Cressie, 1986). Perform universal kriging of the coal ash data as follows:

```
> coal.krige <- krige(coal ~ loc(x, y) + x + x^2,
+        data = coal.ash, covfun = spher.cov,
+        range = 4.31, sill = 0.14, nugget = 0.89)
> coal.predict <- predict(coal.krige)
```

View a summary of the `coal.krige` object as follows:

```
> coal.krige
Call:
krige(formula = coal ~ loc(x, y) + x + x^2,
        data = coal.ash, covfun = spher.cov,
        range = 4.31, sill = 0.14, nugget = 0.89)

Coefficients:
 constant         x          x^2
 9.633667 -1.30365 -0.1383046

Number of observations: 208
```

In addition to a summary of the call, estimates for the coefficients of the fitted surface are displayed.

Plot the surface based on the kriging predictions using the Trellis graphics `wireframe` function:

```
> wireframe(fit ~ x * y, data = coal.predict,
+        screen = list(z = 300, x = -60, y = 0),
+        drape = T)
```

The resulting surface is shown in figure 4.25. A similarly produced wireframe plot of the kriging prediction standard errors, using `se.fit`, is shown in figure 4.26.

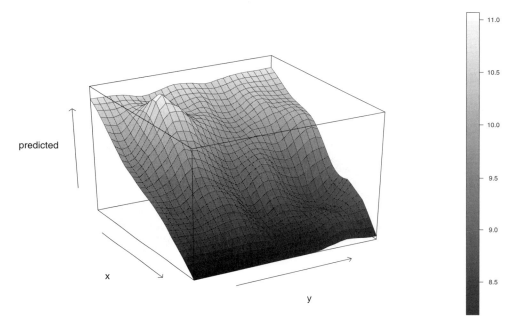

FIGURE 4.25. Surface plot of percent coal ash based on universal kriging predictions.

The modeled trend in the east-west direction is clear from the plot of the predicted surface. The plot of the standard errors reflects the shape of the sampled region; areas on the edges of the prediction grid have higher standard errors since the prediction locations are further away from sampled locations. One might consider kriging over a different prediction region, as was discussed for the scallops data kriging example in section 4.3.1.

4.4 Simulating Geostatistical Data

Simulated data can be useful for examining sampling strategies, as well as for evaluating various prediction methods.

In S+SPATIALSTATS geostatistical data with known autocorrelation can be generated using the function `rfsim`. For example, to simulate a random field on a 20×20 grid, proceed as follows:

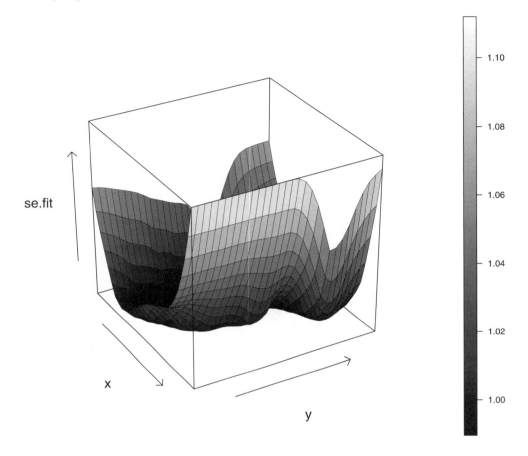

FIGURE 4.26. Surface plot of universal kriging prediction standard errors.

```
> xy20 <- expand.grid(x=seq(0.5,10,len=20),
+       y=seq(0.5,10,len=20))
> z.exp <- rfsim(xy20,covfun=exp.cov,range=2)
```

The function expand.grid returns a 400×2 data frame of the spatial
locations over the grid. The spatial locations are then input into rfsim,
along with a model for the covariance structure. By default, the output is
a Gaussian random field. The values in *z.exp* are autocorrelated with an
exponential distanced based covariance. Figure 4.27 shows the simulated
surface and autocorrelation structure, plotted as follows:

```
> par(mfrow=c(2,1))
> persp(unique(xy20$x),unique(xy20$y),
+       matrix(z.exp,20))
```

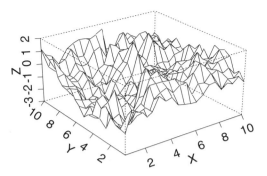

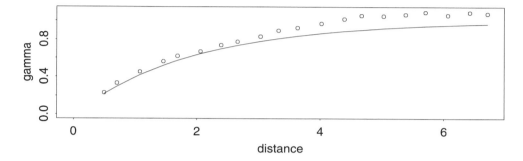

FIGURE 4.27. Perspective plot of the simulated surface (top), the empirical variogram based on the simulated data (open circles, bottom), and the theoretical autocorrelation structure used for the simulation (line, bottom).

```
> z.var <- variogram(z.exp~loc(xy20$x,xy20$y))
> plot(z.var,ylim=c(0,1.1))
> lines(z.var$distance,exp.vgram(
+      z.var$distance,range=2))
```

This simulation represents only one realization of a Gaussian random field with an exponential distanced based covariance.

5

Analyzing Lattice Data

This chapter introduces procedures available in S+SPATIALSTATS for analyzing and modeling lattice data. Lattice data are observations from a random process observed over a countable collection of spatial regions, and supplemented by a neighborhood structure. The observation locations can be regular (equally spaced grid) or irregular, and data at a particular location typically represent the entire region. The data observed at each site may be continuous or discrete. For example, the sample data frame `sids` contains rates of Sudden Infant Death Syndrome (SIDS) for each North Carolina county for the time periods 1974–1978 (Cressie and Chan, 1989). The locations are the coordinates of the county seats. This is discrete data residing on an irregular lattice with each site representing an entire county. For a more rigorous definition of lattice data, see chapter 1.

Before modeling the spatial component of lattice data in S+SPATIALSTATS, we assume *stationarity* (see Glossary for definition) and multivariate normality of the data. This means that trend must be removed, and transformations may be required to stabilize the variance and/or to approximate normality. In section 3.3, basic exploratory data analysis (EDA) techniques were used to check these assumptions on the SIDS data. We will use the results of that preliminary analysis and expand on it throughout this chapter.

In this chapter you will learn to do the following tasks in S+SPATIALSTATS:

- Define spatial neighbors (section 5.1).

- Test lattice data for spatial autocorrelation (section 5.2).

- Model lattice data using spatial regression (section 5.3).

- Simulate lattice data (section 5.4).

5.1   Spatial Neighbors

Lattice modeling is the spatial analogue to time series modeling. A time series is modeled by predicting the outcome for each time based on its dependence on the preceding observation or set of observations (serial autocorrelation). A spatial process is modeled by predicting the outcome for each region based partially on its dependence on nearby or *neighboring* regions. If two regions are neighbors then random processes measured at these regions might be spatially correlated. Choosing a *neighborhood* structure is the first step in the analysis of lattice data. The result determines the covariance structure used for the spatial component of a more general linear regression model.

Neighbors may be defined as regions which border each other, or as regions within a certain distance of each other. For example, in figure 5.1, the county in the center might have only the adjacent (shaded) areas as neighbors. If the correlation structure is defined in terms of distance, the neighbor set might also include the lightly shaded counties.

The neighbor relationship is not necessarily symmetric. For example, the underlying process of interest may flow in only one direction, or a region that is very large might exhibit influence on, but not be influenced by, a smaller region. Since the neighborhood structure is the basic structure for the covariance model for lattice data, the careful definition of spatial neighbors is a crucial analysis step.

5.1.1   Objects of Class "spatial.neighbor"

Lattice modeling in S+SPATIALSTATS requires that neighbor information be held in an object of class "spatial.neighbor". The S-PLUS object sids.neighbor contains the neighbor information for the SIDS data:

```
> sids.neighbor[1:15,]
Total number of spatial units =  100
(Matrix was NOT defined as symmetric)
    row.id col.id    weights matrix
  2      1     17 0.04351368        1
  3      1     19 0.04862620        1
  4      1     32 0.10268062        1
```

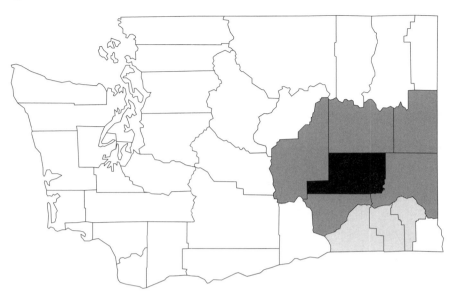

FIGURE 5.1. County map of Washington used to illustrate possible neighbor structure.

5	1	41	0.20782813	1
6	1	68	0.11500900	1
8	2	14	0.17520402	1
9	2	18	0.27140700	1
10	2	49	0.20882988	1
11	2	97	0.22700297	1
13	3	5	0.16797229	1
14	3	86	0.25458569	1
15	3	97	0.22660765	1
17	4	62	0.06865403	1
18	4	77	0.15401887	1
19	4	84	0.09565681	1

The first two columns of an object of class `"spatial.neighbor"` identify the pairs of regions which are neighbors. They form a sparse matrix representation of the spatial neighbor (or contiguity) matrix. The spatial neighbor matrix has non-zero values, given by the `weights` column, for each [row,column] ([`row.id,col.id`]) combination which represents a neighbor pair, and zero everywhere else. For example, the regions indexed by 1 and 17 are neighbors with a neighbor weight of 0.0435. The choice of values for neighbor weights is discussed in section 5.1.4. The `matrix` column allows the user to specify more than one neighbor or neighbor weight configuration in the same spatial neighbor. See section 5.1.5 for an example using two neighbor matrices.

An object of class `"spatial.neighbor"` has two attributes. The total number of spatial units or regions is maintained for sparse matrix routines in attribute `nregion`, since some regions may have no neighbors. The `symmetry` attribute is a logical flag indicating whether or not the neighbor matrix is symmetric. The object `sids.neighbor` was not defined as symmetric, so no assumption of symmetry or lack of symmetry will be made for calculations involving this spatial neighbor object. For more discussion on symmetry, see section 5.1.3.

5.1.2   Reading Neighbor Information from an ASCII File

If your neighborhood information is stored outside of S-PLUS, it may be necessary to import it using the function `read.neighbor`. Since the number of neighbors typically varies between regions, we can not in general use a simple call to the S-PLUS function `scan`. For example, suppose your data are in an ASCII file "**neighbor.dat**" of the form:

```
1 2 3 5
2 1 3 4 5
3 2 5 6 1
4 2 5
5 2 3 4 6 1
6 3 5
```

where each row identifies a region and its neighbors. For example, region 1 has regions 2, 3, and 5 as neighbors, and so on. Each record consists of one fixed element or variable, the region identifier, and a list of neighbors which is free to vary in length. Read this file into S-PLUS as follows:

```
> readnhbr.test <- read.neighbor("neighbor.dat", keep=F)
> readnhbr.test
Total number of spatial units =  6
(Matrix was NOT defined as symmetric)
   row.id col.id weights matrix
 1    1     2      1       1
 2    1     3      1       1
 3    1     5      1       1
 4    2     1      1       1
 5    2     3      1       1
 6    2     4      1       1
 7    2     5      1       1
```

8	3	2	1	1
9	3	5	1	1
10	3	6	1	1
11	3	1	1	1
12	4	2	1	1
13	4	5	1	1
14	5	2	1	1
15	5	3	1	1
16	5	4	1	1
17	5	6	1	1
18	5	1	1	1
19	6	3	1	1
20	6	5	1	1

We have specified `keep=F` to avoid creating an extra component containing only the region identifier. This is adequate in this case, since the regions are simply numbered from 1 to n, the number of regions. If your regions are not numbered like this, you will need to make adjustments to your spatial neighbor, and you might want to retain the region identifiers by using `keep=T`. See section 5.1.3 for information on remapping region identifiers.

The `weights` and `matrix` columns returned by `read.neighbor` default to all values of 1; these may need to be adjusted for model computations.

If the file containing your neighbor information contains other site-specific data, you can still use `read.neighbor`. For example, say you have an ASCII file "**sids.dat**" where the first five fields are site-specific data, and the remaining fields are neighbors for each site:

1	278	151	4672	13	17 19 32 41 68
2	179	142	1333	0	14 18 49 97
3	183	182	487	0	5 86 97
4	240	75	1570	15	62 77 84 90
5	164	176	1091	1	3 95 97
6	138	154	781	0	12 14 56 61 95 100
7	406	118	2692	7	59 74 94
8	411	148	1324	6	21 46 59 94
.					
.					
100	120	142	770	0	6 11 56 58 61

Note: This data file is not included with S+SPATIALSTATS.

Read in this file as follows:

```
> sids.test <- read.neighbor("sids.dat",
+      field.names=c("id","easting", "northing",
```

```
+        "births", "sid"), region.id=1)
> sids.test$data[1:8,]
$data:
  id easting northing births sid
1  1     278      151   4672  13
2  2     179      142   1333   0
3  3     183      182    487   0
4  4     240       75   1570  15
5  5     164      176   1091   1
6  6     138      154    781   0
7  7     406      118   2692   7
8  8     411      148   1324   6
> sids.test$nhbr[1:10,]
Total number of spatial units =  100
(Matrix was NOT defined as symmetric)
   row.id col.id weights matrix
 1      1     17       1      1
 2      1     19       1      1
 3      1     32       1      1
 4      1     41       1      1
 5      1     68       1      1
 6      2     14       1      1
 7      2     18       1      1
 8      2     49       1      1
 9      2     97       1      1
10      3      5       1      1
```

Since this file has data in addition to neighbor information, the default `keep=T` stores the data in the data frame `$data`. The argument `field.names` is required when there are fixed-length records other than the region identifiers. Notice that the variable-length neighbor fields are not named. The argument `region.id` locates the region identifier in the set of field names. The default chooses the last field name (the last variable in the list of fixed variables) for the region identifiers.

The defaults for `read.neighbor` presume that all fields are numeric and that the variable length list of neighbors appears at the end of each record. The arguments `all.numeric`, `char`, and `first.neighbor` allow for different circumstances. See the help file for more information and examples using `read.neighbor`.

5.1.3 Finding Spatial Neighbors

If you have not designated the spatial neighbors prior to loading your data into S-PLUS, there are several functions available to help you do so. If your data were collected on a regular lattice, use `neighbor.grid` to create an object of class `"spatial.neighbor"` for your grid. If your data are located on an irregular lattice, use the function `find.neighbor` instead.

Create an object of class `"spatial.neighbor"` for a 3×3 grid with first-order neighbor relationships as follows:

```
> ng <- neighbor.grid(nrow=3, ncol=3,
+        neighbor.type="first.order")
> ng[1:10,]
Total number of spatial units =  9
(Matrix was NOT defined as symmetric)
   row.id col.id weights matrix
1      1      2       1      1
2      1      4       1      1
3      2      1       1      1
4      2      3       1      1
5      2      5       1      1
6      3      2       1      1
7      3      6       1      1
8      4      1       1      1
9      4      5       1      1
10     4      7       1      1
```

The grid is numbered consecutively down columns, and the `"first.order"` method identifies neighbors as those to the left, to the right, above or below each region (sometimes called the *rook* pattern).

There are four other standard neighbor types available for `neighbor.grid`: `"second.order"` which includes the first-order neighbors and those diagonally linked (the *queen* pattern); `"diagonal"` with neighbors *only* on the diagonal (the *bishop* pattern); and `"hexagonal.in"` and `"hexagonal.out"` for hexagonal grids. See the help file for more information on these patterns. Alternatively, the user can specify a unique neighbor relationship by defining the pattern and weight computation. The following example will form the spatial neighbor for this pattern of neighbor weights:

```
0  .5   0
1   X   1
0  .5   0
```

1. First, write a function that returns the appropriate weight values between the locations (`row1`,`col1`) and (`row2`,`col2`):

```
> my.weight.fun <- function(row1,col1,row2,col2){
+         if(abs(row1 - row2) == 1)
```

```
+                   if(col1 == col2)
+                       return(0.5)
+                   else return(0)
+           else if(row1 == row2)
+                   if(abs(col1 - col2) == 1)
+                       return(1)
+                   else return(0)
+           else return(0)
+ }
```

2. Use `neighbor.grid` with `neighbor.type="user"`, and `weight.fun` given by the function created in step 1:

```
> ng2 <- neighbor.grid(nrow=5, ncol=5,
+       neighbor.type="user", weight.fun=my.weight.fun,
+       max.horiz.dist=1, max.vert.dist=1)
> ng2[1:10,]
Total number of spatial units =  25
(Matrix was NOT defined as symmetric)
```

	row.id	col.id	weights	matrix
1	1	2	0.5	1
2	1	6	1.0	1
3	2	1	0.5	1
4	2	3	0.5	1
5	2	7	1.0	1
6	3	2	0.5	1
7	3	4	0.5	1
8	3	8	1.0	1
9	4	3	0.5	1
10	4	5	0.5	1

The optional arguments `max.horiz.dist` and `max.vert.dist` reduce computation time for large grids by limiting the search region for neighbors.

Both **ng** and **ng2** created above contain symmetric neighbor relationships. However, the attribute **symmetry** of each spatial neighbor has defaulted to **FALSE** ("Matrix was NOT defined as symmetric" appears above the spatial neighbor). Use the S+SPATIALSTATS function `spatial.condense` to change this attribute and remove redundant information from **ng2** as follows:

```
> ng2 <- spatial.condense(ng2, symmetry=T)
> ng2[1:10,]
Total number of spatial units =  25
```

```
        (Matrix defined as symmetric)
          row.id col.id weights matrix
          3       2      1     0.5      1
          6       3      2     0.5      1
          9       4      3     0.5      1
         12       5      4     0.5      1
         14       6      1     1.0      1
         17       7      2     1.0      1
         18       7      6     0.5      1
         21       8      3     1.0      1
         22       8      7     0.5      1
         25       9      4     1.0      1
```

 Warning: If you specify `symmetric=T` when using `spatial.condense`, and your matrix is NOT actually symmetric, some model computations will be wrong.

Use `find.neighbor` on an irregular lattice to find the k nearest neighbors, or all of the neighbors within a given distance. To use `find.neighbor`, you must first make a quad tree, using the function `quad.tree`. A quad tree is a sorted matrix, which provides the most efficient ordering for the nearest neighbor search. We will illustrate this procedure using the SIDS data. For Cressie's (1993, p. 385) analysis of this data set, two counties were considered to be neighbors if their county seats are within 30 miles of each other. Find the neighbors within 30 miles of each other as follows:

```
> sids.place <- cbind(sids$easting,sids$northing)
> sids.quad <- quad.tree(sids.place)
> sids.nhbr <- find.neighbor(x=sids.place,
+       quadtree=sids.quad, max.dist=30)
```

```
> sids.nhbr[1:10,]
     index1 index2 distances
   1      1      1    0.00000
   2      1     41   20.02498
   3      1     17   24.18677
   4      1     68   16.00000
   5      1     32   28.44293
   6      1     19   27.29469
   7      2      2    0.00000
   8      2     97   15.13275
   9      2     49   18.86796
  10      2     14   21.00000
```

Note: This results in one more neighbor pair than those used by Cressie [p. 244, 1993] and stored in the spatial neighbor `sids.neighbor`. The distance between regions 74 and 98 is 29.83, so it should be included. However, to maintain consistency with Cressie's results, we have excluded it from our analysis.

The result of `find.neighbor` is a data frame with three columns containing the neighbor pairs and the distances between them. We will use this information to form an object of class `"spatial.neighbor"` in section 5.1.3.

We have identified each region as a neighbor of itself. Remove this redundant information before proceeding:

```
> sids.nhbr <- sids.nhbr[sids.nhbr[,3] != 0,]
```

Creating an Object of Class `"spatial.neighbor"`

To perform analysis on lattice data in S+SPATIALSTATS, your neighbor information must be stored in an object of class `"spatial.neighbor"`. Create this object from neighbor information stored in S-PLUS using the function `spatial.neighbor` as follows:

```
> sids.snhbr <- spatial.neighbor(row.id=sids.nhbr[,1],
+       col.id=sids.nhbr[,2])
```

```
> sids.snhbr[1:10,]
Total number of spatial units =   100
(Matrix was NOT defined as symmetric)
   row.id col.id weights matrix
    2      1      41       1       1
    3      1      17       1       1
    4      1      68       1       1
    5      1      32       1       1
    6      1      19       1       1
    8      2      97       1       1
    9      2      49       1       1
   10      2      14       1       1
   11      2      18       1       1
   13      3      97       1       1
```

The arguments `row.id` and `col.id` must be the row and column indices of your neighbor matrix, like those returned by `find.neighbor`. If the unique elements of the vector `row.id` are NOT the numbers 1 through the total number of spatial units, one of two things must be true:

1. There are islands, regions with data but no neighbors, in your data set.

2. You are using regions identifiers which are not appropriate indices for a neighbor matrix.

If the attribute **nregion** of your spatial neighbor contains an incorrect number of spatial units for your neighbor matrix, the latter is the case. (This attribute is printed as the number of spatial units in the output shown above.) If so, correct your spatial neighbor using the function **check.islands** with argument **remap=T**. See the help file for more information on **check.islands**.

The **weights** column defaults to 1—equal weights for all neighbors—but it is recommended that these weights be carefully defined. The choice of neighbor weights is discussed in section 5.1.4.

If you do not specify symmetry in your call to **spatial.neighbor**, no assumption of symmetry or lack of symmetry is made—calculations should be correct either way. If you have a large matrix with symmetric neighbor relationships, use the optional attribute **symmetric=T** in your call to **spatial.neighbor** to eliminate some redundancy and speed up calculations in some modeling procedures. In this case, you only need to input one of each set of symmetric neighbor pairs.

 Warning: If you specify **symmetric=T** when you create your spatial neighbor object and your matrix is NOT actually symmetric, some model computations will be wrong.

If you are not sure whether your neighbor matrix is symmetric, use the S-PLUS function **is.Hermitian** to check the symmetry as follows:

```
> is.Hermitian(spatial.weights(sids.snhbr), tol=0)
[1] T
```

The S+SPATIALSTATS function **spatial.weights** turns an object of class **"spatial.neighbor"** into a numeric matrix. The argument **tol=0** is required to check for symmetry. The neighbor relationships for the North Carolina counties in **sids.snhbr** are symmetric. This will change when we add neighbor weights (see section 5.1.4).

We can also use the function **"spatial.neighbor"** for a neighbor matrix in non-sparse notation. For example:

```
> n.mat <- matrix(data=c(0,1,.5,0,1,0,.5,0,0,.5,0,1,
+     0,.5,1,0), nrow=4)
> n.mat
```

```
        [,1] [,2] [,3] [,4]
[1,]   0.0  1.0  0.0  0.0
[2,]   1.0  0.0  0.5  0.5
[3,]   0.5  0.5  0.0  1.0
[4,]   0.0  0.0  1.0  0.0

> spatial.neighbor(neighbor.matrix = n.mat)
Total number of spatial units =  4
(Matrix was NOT defined as symmetric)
  row.id col.id weights matrix
1     2      1     1.0     1
2     3      1     0.5     1
3     1      2     1.0     1
4     3      2     0.5     1
5     2      3     0.5     1
6     4      3     1.0     1
7     2      4     0.5     1
8     3      4     1.0     1
```

5.1.4   Neighbor Weights

The `weights` component of an object of class `"spatial.neighbor"` determines the strength of the correlation between neighbors. There are many ways to assign neighbor weights, depending on the type of spatial application. This specification requires *a priori* knowledge of the *range* and *intensity* of the spatial covariance between regions (Griffith, 1995). Common methods include row-standardization, length of common boundary, and distance functions. *Row-standardization* scales the covariance based on the number of neighbors of each region (in each row of the neighbor matrix). For a region (row) with five neighbors, each neighbor pair (non-zero column) would have a weight of $1/5$. This is a non-symmetric neighbor relationship, since each region (row) has its neighbor weights scaled separately.

 Warning: Although there are some situations when negative neighbor weights may be desirable, the lattice methods in this release of S+SPA-TIALSTATS do not allow for negative neighbor weights.

The choice of weights between neighbors is a crucial analysis step. According to Griffith (1995), an incorrect choice of weights inflates the standard error of the model and biases the correlation estimate, particularly for small sample sizes. Misspecification of the effective range of the correlation is more serious than an incorrect functional form for the correlation relationship. Also, it is better to under-specify the weights matrix (have too few neighbors) than to over-specify it (Griffith, 1995). We recommend that more than one set of neighbor weights be tried so that the sensitivity of model results can be determined.

The neighbor weights in the S-PLUS object `sids.neighbor` are based on the distance-decay correlation function used by Cressie (1993, p.557). The correlation between each pair of neighbors is estimated by:

$$c_{ij} = \begin{cases} \rho(d_{ij}^{-k}/C(k))(n_j/n_i)^{1/2} & j \in N_i \\ 0 & \text{otherwise.} \end{cases}$$

where ρ is a constant to be estimated, d_{ij} is the distance between the two neighbors, n_i is the number of births in county i, $C(k) = \max\{d_{ij}^{-k} : j \in N_i; i = 1, ..., n\}$ (a scaling factor), and N_i is the set of neighbors of region i. The neighbor weights are $c_{ij}\rho$. Cressie tries different values for k, and finds that $k = 1$ yields the best model results. Therefore, the following neighbor weights are contained in `sids.neighbor`:

$$g_{ij} = \begin{cases} (\min\{d_{ij} : i = 1, ..., n\}/d_{ij})(n_j/n_i)^{1/2} & j \in N_i \\ 0 & \text{otherwise.} \end{cases} \tag{5.1}$$

To create these neighbor weights in S-PLUS:

```
> dij <- sids.nhbr[,3]        # distances between neighbors
> n <- sids$births  # births in each county
> el1 <- min(dij)/dij
> el2 <- sqrt(n[sids.snhbr$col.id]/n[sids.snhbr$row.id])
> sids.snhbr$weights <- el1*el2
> sids.snhbr[1:10,]
Total number of spatial units =  100
(Matrix was NOT defined as symmetric)
    row.id col.id      weights matrix
  2      1     41 0.20782813      1
  3      1     17 0.04351368      1
  4      1     68 0.11500900      1
  5      1     32 0.10268062      1
  6      1     19 0.04862620      1
  8      2     97 0.22700297      1
  9      2     49 0.20882988      1
 10      2     14 0.17520402      1
 11      2     18 0.27140700      1
 13      3     97 0.22660765      1
```

The S-PLUS object `sids.neighbor` contains the same information as the object `sids.snhbr` created above. Check the symmetry of `sids.neighbor` as follows:

```
> is.Hermitian(spatial.weights(sids.neighbor),tol=0)
[1] F
```

Our definition of neighbor weights has created an asymmetric neighbor matrix.

5.1.5 Using More Than One Neighbor Matrix

The `matrix` component of an object of class `"spatial.neighbor"` is used to model different types of neighbors separately. For example, we might expect a different correlation level between neighbors 0–20 miles apart and those 20–40 miles apart. Create a new spatial neighbor for the SIDS data to reflect this difference:

1. Find the neighbors within the given distances:

```
> sids.20nhbr <- find.neighbor(x = sids.place,
+       quadtree=sids.quad, max.dist=20)
> sids.40nhbr <- find.neighbor(x = sids.place,
+       quadtree=sids.quad, max.dist=40)
```

2. Remove self-neighbors and redundant information:

```
> sids.20nhbr <- sids.20nhbr[sids.20nhbr[,3] !=0,]
> sids.40nhbr <- sids.40nhbr[sids.40nhbr[,3]>20,]
```

3. Combine the two neighbor lists:

```
> row.id2040 <- c(sids.20nhbr[,1],sids.40nhbr[,1])
> col.id2040 <- c(sids.20nhbr[,2],sids.40nhbr[,2])
```

4. Create the `matrix.id` column, designating the 0–20 mile neighbors as `matrix.id=1`, and the 20–40 mile neighbors as `matrix.id=2`:

```
> dim(sids.20nhbr)
[1] 140   3
> dim(sids.40nhbr)
[1] 590   3
> matrix.id <- c(rep(1,140),rep(2,590))
```

5. Create the spatial neighbor object:

```
> sids.2nhbr <- spatial.neighbor(row.id=row.id2040,
+       col.id=col.id2040, matrix.id=matrix.id)
```

6. Create the distance-decay neighbor weights:

```
> dij2 <- c(sids.20nhbr[,3], sids.40nhbr[,3])
> elem12 <- min(dij2)/dij2
> nvec <- sids$births
> col.id <- sids.2nhbr$col.id
> row.id <- sids.2nhbr$row.id
> elem22 <- sqrt(nvec[col.id]/nvec[row.id])
> sids.2nhbr$weights <- elem12*elem22
```

This spatial neighbor will be used to test spatial autocorrelation in section 5.2, and to fit a spatial regression model in section 5.3.2.

5.2 Spatial Autocorrelation

If a process is spatially autocorrelated, there may be a need for spatial modeling. A test for spatial autocorrelation can be performed as an exploratory technique to decide whether spatial modeling should be used. It can also be useful for testing residuals from a standard multiple regression or trend surface model, but this test is not the preferred method for residuals (Ripley, 1981, p. 99). The null hypothesis is of no correlation, and the alternative hypothesis is specifically defined by a weighted neighbor matrix. The result is therefore sensitive to the choice of neighbors and weights, so it may be desirable to run the autocorrelation under several different scenarios. The calculation of spatial autocorrelation assumes constant mean and variance. If the process contains trend or nonconstant variance, the results should be used with caution.

The S+SPATIALSTATS function `spatial.cor` computes the two common measures of spatial autocorrelation, the Moran and Geary statistics. Another option is a to let the user define an autocorrelation measure. See Cliff and Ord (1981, p. 17) or the help file for formulas and further definition of these measures.

The occurrence of SIDS is not likely to have constant variance, since counties with low birth rates will have more variance. The rates have been transformed using the Freeman-Tukey square root transformation (`sids$sid.ft`) (see section 3.3). Cressie (1993) determined that the square root of the number of births times these transformed rates have approximately equal variances. This is reasonable because it approximates the square-root transformation of a Poisson random variable.

Check the autocorrelation of the 1974–1978 SIDS mortality rates using the Moran statistic and `sids.neighbor` as follows:

```
> sids.cor1 <- spatial.cor(sids$sid.ft*sqrt(sids$births),
+       neighbor=sids.neighbor, statistic="moran",
+       sampling="free", npermutes=1000)
```

The variance can be calculated in either of two ways, depending on the sampling attribute. The "free" method assumes normality—that the data are independent observations from a normal distribution or distributions. The "nonfree" method uses randomization to calculate the variance, without assuming a specific distribution. This corresponds to sampling with replacement ("free") or without replacement ("nonfree"). For the SIDS data, we checked the distribution of the Freeman-Tukey-transformed SIDS rates in section 3.3, and saw no evidence of nonnormality, so the "free" method seems reasonable. Under this assumption, we can use the sample autocorrelation coefficient to test the hypothesis of no autocorrelation versus the alternative of positive spatial autocorrelation.

The optional argument npermutes is used to set the number of random permutations of the data vector on which to calculate the sample autocorrelation. This permutation distribution can then be used to test the significance of the computed coefficient. If npermutes is not given, it defaults to zero.

Display the result as follows:

```
> sids.cor1

Spatial Correlation Estimate

Statistic = "moran" Sampling = "free"

Correlation =  0.2287
Variance    =  0.00772
Std. Error  =  0.08786

Normal statistic =  2.717
Normal p-value (2-sided) =  0.006579

Null Hypothesis:  No spatial autocorrelation

Quantiles of the permutation-correlations :
    Min.  1st Qu. Median    Mean 3rd Qu.   Max.
 -0.3997 -0.06338 -0.011 -0.0114 0.03964 0.4766

permutation p-value = 0.008
```

Under most conditions, the Moran statistic is approximately normally distributed, with mean $-1/(n-1)$ under the null hypothesis, where n is the number of regions. The range and the variance of the statistic are functions of the neighbor weights. To evaluate the significance of the correlation coefficient, compare it to the two-tailed normal p-value or the permutation p-value given in the output above. Both methods yield a significant result for the SIDS data.

Compare the results using the Geary correlation coefficient as follows:

```
> sids.cor2 <- spatial.cor(sids$sid.ft*sqrt(sids$births),
+       neighbor=sids.neighbor, sampling="free",
+       statistic="geary", npermutes=1000)
> sids.cor2

Spatial Correlation Estimate

Statistic = "geary" Sampling = "free"

Correlation =  0.7439
Variance    =  0.0129
Std. Error  =  0.1136

Normal statistic =  -2.255
Normal p-value (2-sided) =  0.02414

Null Hypothesis:  No spatial autocorrelation

Quantiles of the permutation-correlations :
   Min. 1st Qu. Median  Mean 3rd Qu.  Max.
 0.6008  0.9178 0.9913 1.001   1.066 1.739

permutation p-value = 0.011
```

The Geary statistic is also asymptotically normal, with range and variance depending on the neighbor weights. However, there are important differences from the Moran statistic: the mean of the Geary statistic under the null hypothesis is 1, it is never negative, and *low* values (between 0 and 1) indicate *positive* spatial autocorrelation. Thus, the Geary statistic also yields significant positive autocorrelation.

We can also use `spatial.cor` on a spatial neighbor with multiple neighbor matrices. Find the Moran statistics for the two-matrix spatial neighbor created in section 5.1.5 as follows:

```
> spatial.cor(sids$sid.ft*sqrt(sids$births),
+       neighbor=sids.2nhbr, sampling="free",
+       statistic="moran", npermute=1000)

Spatial Correlation Estimate
```

```
Statistic = "moran" Sampling = "free"

Correlation =   0.1995 0.2285
Variance    =   0.02396   0.003349
Std. Error  =   0.1548   0.05787


Normal statistic =   1.354 4.122
Normal p-value (2-sided) =   0.1758    3.753e-5


Null Hypothesis:   No spatial autocorrelation

Quantiles of the permutation-correlations :
              x1.1      x1.2
    Min.  -0.62520 -0.16170
 1st Qu.  -0.09728 -0.04986
  Median  -0.00776 -0.01333
    Mean  -0.01097 -0.01196
 3rd Qu.   0.07841  0.02323
    Max.   0.52250  0.19920


permutation p-value = 0.069 0.000
```

The second neighbor matrix, which contains neighbors defined as those counties within 20–40 miles of each other, has significant positive spatial autocorrelation. The 0–20 mile neighbor matrix has borderline significant spatial autocorrelation. This may be the effect of a relatively low number of neighbors (140) in the 0–20 matrix.

Results of spatial autocorrelation tests must be used with caution. First, the choice of neighbors and their respective weights determine the values of the Moran and Geary statistics. A non-significant result indicates that there is no significant spatial autocorrelation following the provided neighbor structure. Griffith (1995) reports that misspecification of the neighbor weights causes a loss in statistical power in the test for spatial autocorrelation, with the Moran statistic yielding slightly better power. If there are other possibilities for the neighbor relationships, rerun the correlation tests under these possibilities to avoid missing patterns at different scales. Second, a significant positive autocorrelation could be caused by a trend in the data. From our exploratory data analysis of the SIDS data in section 3.3, colored probability maps revealed some spatial trend. In this case, a regression model based on spatial location in conjunction with an autoregressive covariance model might be appropriate.

See Haining (1990) or Cliff and Ord (1981) for more discussion on the choice of autocorrelation statistics and sampling methods.

5.3   Spatial Regression Models

To model spatial lattices, we look at two levels of variation—large-scale changes in the mean due to spatial location or other explanatory variables, and small-scale variation due to interactions with neighbors. S+SPATIAL-STATS contains functions to fit spatial regression models of the following form:

$$Z_i = \mu_i + \delta$$

where Z_i is the random process at site i; μ_i is the mean at site i, which may be constant or a linear model with covariates; $\delta \sim N(0, \Sigma)$; and Σ is the covariance matrix of random variables at all sites. Non-constant μ can be modeled as a linear model within the spatial modeling framework of S+SPATIALSTATS. The small-scale variation is modeled by fitting an autoregressive or moving average covariance model to Σ. The parameters of these two model portions interact, and the model is fitted iteratively.

5.3.1   Covariance Families

There are three possible choices for covariance structures available in S+SPATIALSTATS, corresponding to conditional spatial autoregression (CAR), simultaneous spatial autoregression (SAR), and moving average (MA) models. All of these models assume multivariate normality, with the differences lying in the choice of dispersion matrix.

The three dispersion matrix models are:

$$CAR: \quad \Sigma = (I - \rho N)^{-1} D \sigma^2$$

$$SAR: \quad \Sigma = [(I - \rho N)^T D^{-1} (I - \rho N)]^{-1} \sigma^2$$

$$MA: \quad \Sigma = (I + \rho N) D (I + \rho N)^T \sigma^2$$

where ρ and σ are scalar parameters to be estimated by spatial regression, N is a weighted neighbor matrix, and D is a diagonal matrix used to account for nonhomogeneous variance of the marginal distributions.

Note: The matrix D is sometimes called a weights matrix. This is not to be confused with the neighbor weights which are the elements of N in the formulas above. In this document, we refer to the elements of N as neighbor weights.

Both the CAR and SAR models correspond to autoregressive procedures in time series analysis. The residuals from the SAR model are correlated with the neighbor data values, resulting in inconsistent least-squares parameter estimates (Cressie, 1993, p.408). The CAR model does not have this problem, so it is preferred as long as the dispersion matrix is symmetric, which is required. In some cases, as we shall see with the SIDS data, the matrix D can be used to symmetrize the dispersion matrix. The MA model is analogous to the moving average time series model. For further discussion of the appropriate use of CAR, SAR, and MA models, see Cliff and Ord (1981), Cressie (1993), or Haining (1990).

5.3.2   Fitting Spatial Linear Models

The S+SPATIALSTATS function `slm` fits a linear model with spatial dependence, using generalized least squares regression. It works similarly to the S-PLUS functions `lm` and `glm`, with the addition of spatial covariance modeling.

The large-scale or linear model component of `slm` can be a simple linear model, or it can include a *trend surface model* to model the trend using a polynomial based on data locations. For example, a second-order trend surface model with data locations represented by (x_i, y_i) takes the following form:

$$\mu_i = \beta_{10}x_i + \beta_{20}x_i^2 + \beta_{11}x_iy_i + \beta_{01}y_i + \beta_{02}y_i^2,$$

where the βs are the coefficients to be estimated. This can be accomplished with `slm` by using the appropriate functions of the location vectors as explanatory variables in the model formula.

Rather than fitting μ within `slm`, you could model the mean using other functions in S-PLUS, then model the remaining correlated errors using `slm`. For instance, this would be necessary to implement a spatially lagged model. A spatially lagged model is one in which the trend or mean value in each region is a weighted average of the mean value in neighboring regions. The explanatory variable is the row-standardized weighted neighbor matrix multiplied by the variable of interest. This variable of interest might be the dependent variable itself, or another independent variable which affects the process.

For the SIDS data set, we will look at four simple models for large-scale variation in the Freeman-Tukey-transformed SIDS rates: the *null* model (assuming a constant mean); a *race* model using a linear predictor (non-white birth rate) for the mean; a *group* model with 12 means corresponding to clustered regions; and the race model run with the two neighbor matrices formed in section 5.1.5. The model fits will be compared in section 5.3.3.

We would like to use the CAR model for the covariance, but the neighbor weights we have chosen are not symmetric (see section 5.1.4). Counties with higher birth rates exert a stronger influence on their neighbors than counties with lower birth rates. Recall from section 3.3 that the conditional variances of SIDS rates are also dependent on birth rate:

$$var(Z_i|Z_j : j \in N_i) = \sigma_i{}^2$$
$$= \sigma^2/n_i.$$

If we use the diagonal matrix D, with $d_i = 1/n_i$, it will simultaneously correct the variance and symmetrize the covariance matrix. To see that the covariance matrix will be symmetric, first note from equation 5.1 that:

$$g_{ij} * n_i = g_{ji} * n_j$$

This shows that the matrix $(D^{-1} * N)$ is symmetric, and therefore the CAR covariance matrix is also symmetric.

To estimate the parameters for the null linear model using CAR on the SIDS data for 1974–1978:

```
> sids.nullslm <- slm(sid.ft ~ 1, cov.family = CAR,
+       data = sids, subset= -4, spatial.arglist=
+       list(neighbor=sids.neighbor, region.id = 1:100,
+       weights = 1/sids$births))
```

To use `slm`, you must first provide a formula for the linear model. If you have already removed the mean, use the formula $y \sim -1$ to fix the mean at zero. Next, provide your choice of covariance family, and `spatial.arglist`, a list of arguments required to fit the covariance model. In its most basic form, `spatial.arglist` consists only of the spatial neighbor. The optional `subset` argument can be used to remove outliers which might exhibit undue influence on the model fit. For the SIDS data, we remove county 4, which was identified as an outlier in section 3.3. When `subset` is used, you must also supply the argument `region.id` within `spatial.arglist` to maintain the total number of regions. The optional `weights` argument is the diagonal matrix D as discussed above.

View the results of the model fit as follows:

```
> sids.nullslm
Call:
slm(formula = sid.ft ~ 1, cov.family = CAR, data = sids,
subset = -4, spatial.arglist = list(neighbor =
sids.neighbor, region.id = 1:100,
        weights = 1/sids$births))

Coefficients:
 (Intercept)
    2.839042
```

```
Degrees of freedom: 99 total; 97 residual

sigma^2 =  1442.893
rho =  0.8334194

Iterations =  12
Gradient norm =  1.014761e-05
Log-likelihood =  -211.8542
Convergence:  RELATIVE FUNCTION CONVERGENCE
```

The coefficient estimate (the grand mean) is 2.839. The parameter estimates are 1442.893 for σ^2 and 0.8334 for ρ. The log-likelihood for this model is -211.8542.

In section 3.3 we explored the relationship between race and SIDS rates by looking for association between the non-white birth rates and the SIDS rates in North Carolina (see figure 3.31). Since there does appear to be some positive association, fit the spatial regression model with transformed non-white birth rates as a linear explanatory variable:

```
> sids.raceslm <- slm(sid.ft ~ nwbirths.ft,
+       cov.family = CAR, data = sids, subset = -4,
+       spatial.arglist = list(neighbor = sids.neighbor,
+       region.id = 1:100, weights = 1/sids$births))

> summary(sids.raceslm)
 Call:
slm(formula = sid.ft ~ nwbirths.ft, cov.family = CAR,
data = sids, subset = -4, spatial.arglist =
list(neighbor =sids.neighbor, region.id = 1:100,
        weights = 1/sids$births))
Residuals:
  Min    1Q Median    3Q  Max
 -106 -18.79   7.01 26.27 77.8

Coefficients:
            Value Std. Error t value Pr(>|t|)
(Intercept) 1.6456 0.2361     6.9698  0.0000
nwbirths.ft 0.0345 0.0066     5.2602  0.0000

Residual standard error: 34.1745 on 96 degrees of freedom

Variance-Covariance Matrix of Coefficients
            (Intercept)   nwbirths.ft
(Intercept)  0.05574684 -1.485340e-03
nwbirths.ft -0.00148534  4.313787e-05
```

```
        Correlation of Coefficient Estimates
                  (Intercept) nwbirths.ft
    (Intercept)    1.0000000   -0.9578252
    nwbirths.ft   -0.9578252    1.0000000

    rho =   0.6454

    Iterations =  9
    Gradient norm =   4.422e-7
    Log-likelihood =   -200.4

    Convergence:   RELATIVE FUNCTION CONVERGENCE
```

The generic **summary** function is used above to provide more information
on the coefficient estimates. The coefficient estimates for the linear component are 1.646 and 0.035. The standard errors for the coefficients are
given, along with the associated t-statistic. Both coefficients are significantly different from zero, according to the two-tailed p-values given. The
residual distribution shown is of the scaled residuals, which will be discussed in section 5.3.3. The parameter estimate for ρ has decreased, and
the log-likelihood has increased from the null model.

In Cressie's (1993) analysis of the SIDS data, he identified 12 groups of
contiguous counties with similar SIDS rates. We can model the large-scale
trend of the SIDS rates by using 12 separate means for the groups identified
by the factor sids$group:

```
> sids.gpslm <- slm(sid.ft ~ group -1, cov.family = CAR,
+       data = sids, subset = -4, spatial.arglist =
+       list(neighbor=sids.neighbor, region.id = 1:100,
+       weights = 1/sids$births))
> sids.gpslm
Call:
slm(formula = sid.ft ~ group - 1, cov.family = CAR,
data = sids, subset = -4, spatial.arglist =
list(neighbor=sids.neighbor, region.id=1:100,
        weights = 1/sids$births))

Coefficients:
    group1    group2    group3    group4    group5   group6
  2.054726  2.869104  4.253104  2.476255  2.148905  2.63754
    group7    group8    group9   group10   group11  group12
  3.276925  3.110909  2.678222  2.836351  3.178222  3.685853
```

```
Degrees of freedom: 99 total; 86 residual

sigma^2 =  864.7263
rho =  0.7093103

Iterations =  11
Gradient norm =  2.055277e-06
Log-likelihood =  -185.7641
Convergence:  RELATIVE FUNCTION CONVERGENCE
```

The coefficient estimates are effect sizes for the 12 county groups.

In section 5.1.5 we created a spatial neighbor with two matrices, one for regions within 20 miles of each other (N_1), and one for neighbors between 20 and 40 miles (N_2). In section 5.2, we found two Moran spatial autocorrelation statistics for this spatial neighbor ($\widehat{\rho_1}$ and $\widehat{\rho_2}$). Fit the race model with covariance matrix $\rho_1 N_1 + \rho_2 N_2$ as follows:

```
> sids.race2slm <- slm(sid.ft~nwbirths.ft, cov.family=CAR,
+       data = sids, subset = -4, spatial.arglist =
+       list(neighbor=sids.2nhbr, region.id = 1:100,
+       weights = 1/sids$births))
> summary(sids.race2slm)
Call:
slm(formula = sid.ft ~ nwbirths.ft, cov.family = CAR,
data = sids, subset = -4, spatial.arglist =
list(neighbor = sids.2nhbr, region.id = 1:100,
        weights = 1/sids$births))
Residuals:
    Min    1Q Median    3Q    Max
 -105.1 -18.52  4.616 27.18 83.49


Coefficients:
             Value Std. Error t value Pr(>|t|)
(Intercept) 1.7207 0.2691      6.3947  0.0000
nwbirths.ft 0.0324 0.0073      4.4111  0.0000

Residual standard error: 33.8152 on 95 degrees of freedom

Variance-Covariance Matrix of Coefficients
              (Intercept)    nwbirths.ft
(Intercept)  0.072403377 -1.879554e-03
nwbirths.ft -0.001879554  5.380981e-05

Correlation of Coefficient Estimates
```

```
              (Intercept) nwbirths.ft
  (Intercept)   1.0000000  -0.9522362
  nwbirths.ft  -0.9522362   1.0000000


  rho =  0.3445 1.036


  Iterations =  6
  Gradient norm =  8.274e-7
  Log-likelihood =  -199.9


  Convergence:  RELATIVE FUNCTION CONVERGENCE
```

The linear coefficient estimates have changed, and there are now two estimates for ρ, one for each neighbor matrix.

5.3.3   Model Selection

As in most modeling situations, there is more than one possible model for the SIDS data. Adding a spatial covariance component increases the number of possible variations. Along with the usual problem of overlooking or mis-specifying explanatory variables, the choice of neighbors, neighbor weights, and covariance family also significantly affects the model outcome. In this section, we briefly review some model comparison and residual diagnostic techniques available in S-PLUS and S+SPATIALSTATS.

Compare the nested models fitted in section 5.3.2 (race model versus null model) by performing a likelihood ratio test. The test statistic given by Cressie (1993, p. 562) is:

$$U^2 = 2 * \{(n - p - r)/n\} * (L_p - L_{p+r})$$

where n is the number of samples, p is the number of parameters estimated for the smaller nested model, r is the number of additional parameters in the larger model, L_p is the negative log-likelihood for the smaller nested model, and L_{p+r} is the negative log-likelihood for the larger model. This statistic has asymptotic χ^2 distribution with r degrees of freedom. For the null model versus the race model:

```
> sids.nullslm$objective
[1] 211.8542
> sids.raceslm$objective
[1] 200.402
> U2 <- (2*(99-3-1)/99)*(211.854-200.402)
> 1-pchisq(U2, df=1)
[1] 2.75709e-06
```

The linear model with non-white birth rate as an explanatory variable provides a significant improvement over the null model.

The model coefficients can be tested with the t-statistic given in the slm summary. Alternatively, a likelihood ratio test can be performed to test any of the model coefficients or parameters against hypothesized fixed values, using the function lrt. For example, to test if the linear coefficient for the race model is significantly different from zero:

```
> coef(sids.raceslm)
 (Intercept) nwbirths.ft
    1.645617  0.03454844
> lrt(sids.raceslm, coefficients=c("nwbirths.ft"= 0))
Likelihood Ratio Test

Chisquare statistic = 22.90435, df =1,
p.value = 1.702664e-06

parameters:
 0.8334194

coefficients:
 (Intercept) nwbirths.ft(fixed)
    2.839042                  0
```

The function coef yields the names and estimated values for the coefficients in the specified model. The argument coefficients in lrt specifies that we wish to test the hypothesis that the coefficient named nwbirths.ft is equal to zero. The test result is significant. The parameters and coefficients given by the lrt print method are those for the null hypothesis model.

Similarly, we can check whether the ρ estimate for the neighbor matrix for neighbors between 20 and 40 miles is significantly different from zero:

```
> lrt(sids.race2slm, parameters=c(NA,0))
Likelihood Ratio Test

Chisquare statistic = 2.77897, df =1, p.value = 0.0955096

parameters:
           (fixed)
 0.4635494        0

coefficients:
 (Intercept) nwbirths.ft
    1.602641  0.03580331
```

The parameter estimate is not significantly different from zero (at a 0.05 level of significance). The parameters and the coefficients for the reduced model are given in the output above.

5.3.4 Model Diagnostics

The residuals from an `slm` fit are derived differently for each covariance model (see the help files for `CAR`, `SAR`, and `MA`), but in each case they should be approximately independent normal with constant variance. In the case of the SIDS model, we have used the diagonal weights matrix to stabilize the variance, so the residuals have been scaled by these weights. Begin diagnostics with a visual check for normality:

```
> par(mfrow=c(1,2))
> par(pty="s")
> hist(sids.raceslm$residuals)
> qqnorm(sids.raceslm$residuals)
> qqline(sids.raceslm$residuals)
```

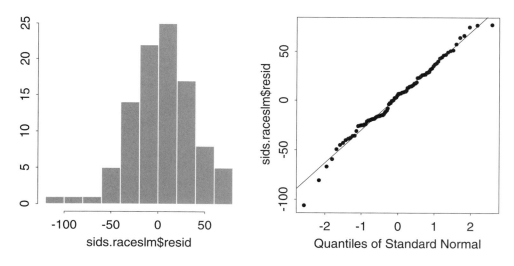

FIGURE 5.2. Histogram and QQ-plots of the residuals from the race model.

The normality in the residuals is marred by some large negative values; see figure 5.2.

Next, plot the residuals against the fitted values to check for homogeneity and outliers:

```
> par(mfrow=c(1,1))
> plot(fitted(sids.raceslm), residuals(sids.raceslm))
> abline(h=0)
> identify(fitted(sids.raceslm), residuals(sids.raceslm),
+       label=sids$id[-4])
[1] 91 33 40
```

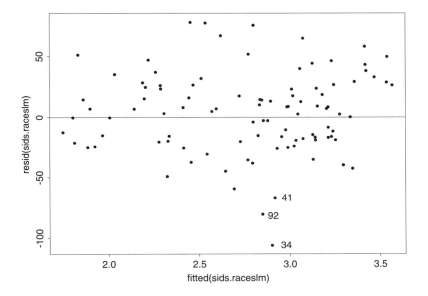

FIGURE 5.3. Residual diagnostics for the race model: residuals vs. fitted values.

The values returned by the function `fitted` are the results of the linear model fit only. The function `residuals` returns the residuals contained in `sids.raceslm$residuals`. We have used the function `identify` to label the large negative residuals in figure 5.3 with their region identifiers. Recall that region 4 has not been included in the model, so the residuals and fitted values have length 99, while the `sids` data frame has 100 records. Look at the data for these regions as follows:

```
> sids[c(34,41,92),]
           id easting northing sid births nwbirths group
Forsyth   34     233      153  10  11858     3919     1
Guilford  41     258      150  23  16184     5483     2
Wake      92     319      133  16  14484     4397     6
            sid.ft nwbirths.ft
Forsyth   1.881463    36.36132
Guilford  2.409886    36.81425
Wake      2.134410    34.84887
```

```
> summary(sids$births)
 Min. 1st Qu. Median Mean 3rd Qu.  Max.
  248    1077   2180 3300    3936 21590
```

All three regions have very high birth rates. The large negative values for these residuals may be due to the scaling: the residuals have been divided by 1/sids$births. However, there are three other counties with birth rates greater than 10,000 which do not have such large residuals.

Look at the data for the neighbors of these regions as follows:

```
> sids[sids.neighbor$col.id[sids.neighbor$row.id==34],]
          id easting northing sid births nwbirths group
Davidson  29     231      133   8   5509      736     5
   Davie  30     213      139   1   1207      148     5
Guilford  41     258      150  23  16184     5483     2
  Stokes  85     233      175   1   1612      160     1
  Yadkin  99     208      156   1   1269       65     1
            sid.ft nwbirths.ft
Davidson  2.483219    23.12491
   Davie  2.197465    22.18395
Guilford  2.409886    36.81425
  Stokes  1.901486    19.95650
  Yadkin  2.143112    14.36867

> sids[sids.neighbor$col.id[sids.neighbor$row.id==41],]
            id easting northing sid births nwbirths group
  Alamance   1     278      151  13   4672     1243     2
   Forsyth  34     233      153  10  11858     3919     1
  Randolph  76     256      126   7   4456      384     6
Rockingham  79     257      173  16   4449     1243     2
              sid.ft nwbirths.ft
  Alamance  3.399155    32.62883
   Forsyth  1.881463    36.36132
  Randolph  2.593262    18.57828
Rockingham  3.851154    33.43657

> sids[sids.neighbor$col.id[sids.neighbor$row.id==92],]
          id easting northing sid births nwbirths group
 Chatham  19     291      127   2   1646      591     6
  Durham  32     306      146  16   7970     3732     2
Franklin  35     337      155   2   1399      736     2
 Harnett  43     309      105   6   3776     1051     6
Johnston  51     335      114   6   3999     1165     6
```

```
          sid.ft nwbirths.ft
 Chatham 2.452338    37.91337
  Durham 2.877352    43.28134
Franklin 2.660029    45.88888
 Harnett 2.622097    33.37480
Johnston 2.547939    34.14369
```

More investigation into these residuals is needed.

Next, plot the residuals against `sids$group` to see if there is any correlation with this potential covariate:

```
> plot(sids$group[-4],residuals(sids.raceslm))
> abline(h=0)
```

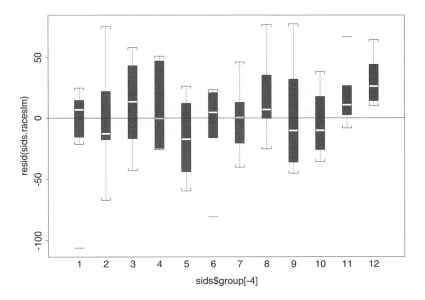

FIGURE 5.4. Residual diagnostics for the race model: residuals vs. group.

There doesn't appear to be a functional relationship in figure 5.4. However, groups 11 and 12 (and perhaps 8) seem to have higher than expected residuals. Plot the locations of the groups to see if they may be a factor:

```
> plot(sids$easting, sids$northing, type="n")
> text(sids$easting, sids$northing, sids$group)
```

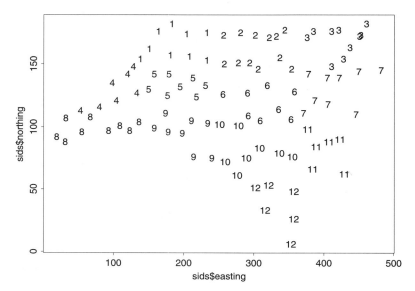

FIGURE 5.5. Spatial locations by **group**.

Figure 5.5 shows that these 3 groups are located along the boundary of the state—an arbitrary human boundary. It is likely that we are seeing the result of edge effects in the residuals of the model.

For the race model, we have modeled the variation in the SIDS data with three components: the linear race component (`fitted(sids.raceslm)`), the neighbor covariance component (termed the signal (Haining, 1990, p. 258)), and the residual variation or noise. The easiest way to get to the signal component for the race model is to subtract the fitted values and the noise from the original data:

```
> noise <- (1/sqrt(sids$births[-4])) * resid(sids.raceslm)
> signal <- sids$sid.ft[-4]-fitted(sids.raceslm)-noise
```

In order for the three components to add to the data, the noise must be the unscaled noise. The signal can also be calculated directly. For the CAR model it is $\rho N(Y - X\beta)$:

```
> nhbr2 <- spatial.subset(sids.neighbor,
+      region.id=c(1:3,5:100), subset=c(1:3,5:100))
> rhoN <- spatial.weights(nhbr2$neighbor,parameter=0.6454,
+      region.id=c(1:3,5:100))
> signal2 <- rhoN %*%
+      (sids$sid.ft[-4] - fitted(sids.raceslm))
```

```
> summary(signal - signal2)
      Min.     1st Qu.      Median      Mean
 -1.617e-05 -2.968e-06 -1.092e-07 0.001053
   3rd Qu.     Max.
 2.083e-06 0.04893
```

The function `spatial.subset` is used to remove region 4, and the function `spatial.weights` computes the matrix of neighbor correlations. The two computations of signal are in reasonable accord.

Predictions from the model can be computed as fitted values + signal. To evaluate the fit of the race model, plot these predictions against the actual SIDS rates:

```
> plot(sids$sid.ft[-4], fitted(sids.raceslm) + signal,
+       ylim=c(1,5.2), xlim=c(1,5.2))
> abline(0,1)
```

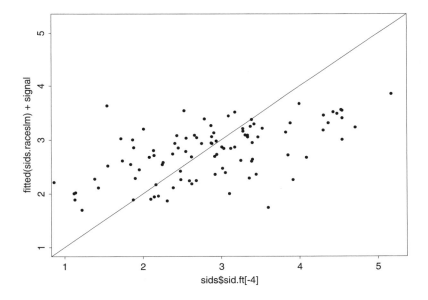

FIGURE 5.6. Actual transformed SIDS rates plotted against those fitted by the race model.

The plot in figure 5.6 shows that the variation in SIDS rates is not completely captured by the race model.

A visual interpretation of the model is possible with colored maps showing predictions versus the data, as follows:

1. Break the SIDS data, the fitted values, and the model predictions into four groups for map coloring:

```
> breaks.sids <- c(-.001, 2.2, 3.0, 3.5, 7)
> sids.y <- c(sids$sid.ft[1:26],
+       rep(sids$sid.ft[27], 3), sids$sid.ft[28:100])
> sids.ygrp <- cut(sids.y, breaks.sids)
> fit2 <- fitted(sids.raceslm)
> fit2 <- c(fit2[1:3], NA, fit2[4:25],
+       rep(fit2[26], 3), fit2[27:99])
> sids.fgrp <- cut(fit2, breaks.sids)
> pred2 <- signal + fitted(sids.raceslm)
> pred2 <- c(pred2[1:3], NA, pred2[4:25],
+       rep(pred2[26], 3), pred2[27:99])
> sids.pgrp <- cut(pred2, breaks.sids)
```

It is necessary to expand the values out to match the number of polygons displayed in the county map. (The 27th county, Currituck, has three polygons associated with it).

2. Create colored state maps:

```
> library(maps)
Warning messages:
  The functions and datasets in library section maps
        are not supported by StatSci. in: library(maps)
> par(mfrow=c(3,1))
> map("county", "north carolina", fill = T,
+       color=sids.ygrp)
> map("county", "north carolina", add=T)
> title(main="Actual Transformed SIDS Rates")
> legend(locator(1), legend=c("< 2.2", "2.2-3.0",
+       "3.0-3.5", "> 3.5"), fill=c(1:4))
> map("county", "north carolina", fill=T,
+       color=sids.fgrp)
> map("county", "north carolina", add=T)
> title(main="Fitted Values")
> legend(locator(1), legend=c("< 2.2", "2.2-3.0",
+       "3.0-3.5", "> 3.5"), fill=c(1:4))
> map("county", "north carolina", fill=T,
+       color=sids.pgrp)
> map("county", "north carolina", add=T)
> title(main="Predictions")
> legend(locator(1), legend=c("< 2.2", "2.2-3.0",
+       "3.0-3.5", "> 3.5"), fill=c(1:4))
```

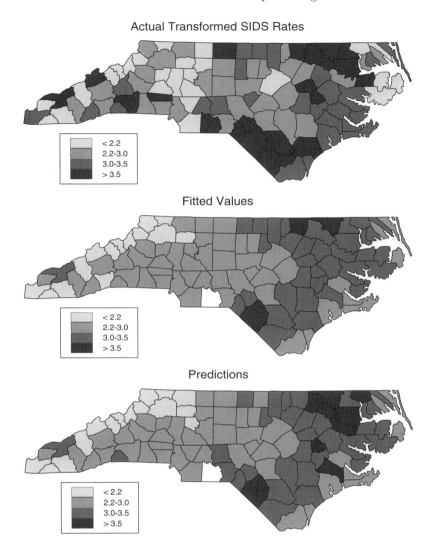

FIGURE 5.7. Colored maps of transformed SIDS rates and the model predictions for the rates. The middle map is for linear race model only, the bottom map includes the spatial component.

The top map in figure 5.7 shows the Freeman-Tukey transformed SIDS rates, which can be compared visually with the predictions mapped in middle and lower maps in the figure. The model predictions shown in the bottom map include the effects of spatial neighbors on the SIDS rates. It is again apparent that the race model has accounted for some, but not all of the spatial variation. Other linear models should therefore be considered, along with different definitions of neighbors and neighbor weights.

5.4   Simulating Lattice Data

Simulated data is useful for validating and examining the properties of the regression models developed in this chapter. This is especially true for spatial modeling, where there are likely to be several models which produce an acceptable fit to our data.

First, a vector of data with mean 0 and covariance Σ is needed. This vector is $y = Le$, where $LL^T = \Sigma$, and $e \sim N(0, 1)$. L is the lower triangular matrix from the Cholesky decomposition of Σ (Haining, 1990, p. 116). Generate one random realization from the SIDS race model with CAR dispersion matrix as follows:

1. Create the diagonal weight matrix, an identity matrix, and ρN:

   ```
   > sids.diag <- diag(1/sids$births)
   > Id <- diag(rep(1,100))
   > rhoN <- spatial.weights(sids.neighbor,
   +       parameters=c(0.6453))
   ```

2. Calculate the CAR covariance matrix, Σ:

   ```
   > sids.carcov <- solve(Id-rhoN) %*% sids.diag*1167.897
   ```

3. Calculate the Cholesky decomposition of Σ:

   ```
   > sids.carcovL <- chol((sids.carcov+t(sids.carcov))/2)
   ```

4. Generate the spatially correlated random vector:

   ```
   > e.norm <- rnorm(100)
   > sids.carsim <- t(sids.carcovL) %*% e.norm
   ```

The function `chol` returns the upper triangular Cholesky matrix, so the transpose is taken to get the lower triangular matrix.

For a SAR model with parameters from a SAR fit of the race model (example not shown), form a correlated vector of errors as follows:

```
> sids.wN2 <- spatial.weights(sids.neighbor,
+       parameters=c(0.4738))
> sids.sarcov <- solve(t(Id-sids.wN2) %*% solve(sids.diag)
+       %*% (Id-sids.wN2))*1159.813
> sids.sarcovL <- chol((sids.sarcov + t(sids.sarcov))/2)
> sids.sarsim <- t(sids.sarcovL)%*%e.norm
```

For a MA model with parameters from an MA fit of the race model (example not shown), form a correlated vector of errors as follows:

```
> sids.wN3 <- spatial.weights(sids.neighbor,
+       parameters=c(0.6337))
> sids.macov <- (Id+sids.wN3) %*% sids.diag
+       %*% t(Id+sids.wN3)*1198.683
> sids.macovL <- chol((sids.macov + t(sids.macov))/2)
> sids.masim <- sids.macovL%*%e.norm
```

For all three covariance models, these simulated errors can be added to an appropriate mean vector to make predictions for the model.

6

Analyzing Spatial Point Patterns

This chapter introduces procedures available in S+SPATIALSTATS for the analysis and modeling of mapped spatial point patterns. A spatial point pattern is a collection of points irregularly located within a bounded region of space. The points can denote locations of naturally occurring phenomena such as earthquakes or plants, or social events such as the locations of small towns or the occurrences of a particular disease. The data set may consist of locations only, or it may be a *marked* point process, with data values associated with each location (marks). An example of a marked point process is a set of tree locations in a forest, along with their associated diameters at breast height.

In section 3.4 some introductory data explorations were performed on the bramble cane data. In this chapter, a second data set containing mapped locations of maple and hickory trees in a 19.6 acre square plot in Lansing Woods, Clinton County, Michigan, will be used for most analyses [(Diggle, 1983, p. 27), (Gerrard, 1969)]. The data have been scaled so that they reside on the unit square, although this is not necessary for analysis using S+SPATIALSTATS.

In this chapter you will learn to do the following tasks in S+SPATIALSTATS:

- Examine point pattern data for complete spatial randomness (section 6.2).

- Estimate the intensity of a spatial point pattern (section 6.3.1).

- Calculate Ripley's K-functions (section 6.3.2).

- Simulate a spatial point process (section 6.4).

6.1   Objects of Class "spp"

A spatial point pattern in this release of S+SPATIALSTATS is a mapped pattern. The coordinates which represent this mapping are stored in objects of class "spp". A point pattern object is a data frame with two columns (typically the first two) containing the locations of the points. More columns may or may not be present. These locational coordinates may be latitude and longitude, easting and northing, or the user's own definition of coordinates. Create a point pattern object from the lansing data frame as follows:

```
> summary(lansing)
      x                    y                 species
 Min.   :0.0010   Min.   :0.0000   hickory:703
 1st Qu.:0.2480   1st Qu.:0.2410   maple  :514
 Median :0.5130   Median :0.5220
 Mean   :0.5094   Mean   :0.5027
 3rd Qu.:0.7720   3rd Qu.:0.7370
 Max.   :1.0000   Max.   :0.9910
```

```
> lansing.spp <- spp(lansing)
```

The function spp creates an object of class "spp" from a matrix or data frame containing locational coordinates x and y. The summary method for a point pattern object shows the total number of points and extent of the coordinates:

```
> summary(lansing.spp)
Total number of points: 1217
Coordinate extents :
 x :  0.001 , 1.000
 y :  0.000 , 0.991
Other covariates :
     species
 hickory:703
 maple  :514
```

The plot method for point pattern objects scales the axes for geometric accuracy. Plot the Lansing data as follows:

```
> par(pty="s")
> plot(lansing.spp, boundary=T)
```

We have used the optional argument boundary=T to plot the rectangular boundary attribute of the Lansing point pattern in figure 6.1. Since

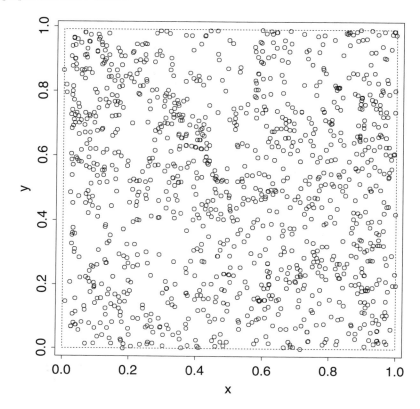

FIGURE 6.1. Scatter plot of the Lansing Woods tree locations.

the boundary was not specified in our call to spp, it has defaulted to the extent of the two locational coordinates found by the S-Plus function scom@bbox@:

```
> bbox(lansing)
$x:
[1] 0.001 0.001 1.000 1.000

$y:
[1] 0.991 0.000 0.000 0.991

> attributes(lansing.spp)$boundary
$x:
[1] 0.001 0.001 1.000 1.000

$y:
[1] 0.991 0.000 0.000 0.991
```

Since the study region is really the unit square, we may want to designate the exact boundary in the call to spp:

```
> lansing.spp <- spp(lansing, boundary=bbox(x=c(0,1),
+       y=c(0,1)))
```

Here bbox computes the bounding box given the limits for x and y.

In general, the boundary of a point pattern object can be any convex polygon. In section 3.4, there is an example of the use of chull for creating the minimum convex polygon that contains the bramble point pattern. Another method is to interactively define the polygon using the function locator. Create a polygon boundary for the maple trees in the Lansing Woods as follows (an example is shown in figure 6.2):

1. Plot the locations of the maple trees with extra space on the plot:

   ```
   > attach(lansing)
   > scaled.plot(x[species=="maple"],
   +       y[species=="maple"], xlim=c(-.2,1.2),
   +       ylim=c(-.2,1.2))
   ```

2. Use locator to choose the boundary, by clicking the left mouse button in the chosen locations on the graphics device. When finished, click the center or right mouse button on the graphics device to exit:

   ```
   > maple.poly <- locator(type="line")
   ```

   Notice that it is not necessary to close the polygon.

3. Check if the polygon is convex:

   ```
   > is.convex.poly(maple.poly)
   [1] T
   ```

The function locator returns the coordinates for the points interactively located on the graphics screen. To test if the boundary is convex, use the function is.convex.poly. The polygon maple.poly can be given as the boundary argument in a call to spp.

6.2 Measures of Spatial Randomness

In this section, we begin the analysis of spatial point patterns with tests for complete spatial randomness (CSR). We define CSR by the following criteria (Diggle, 1983, p. 4):

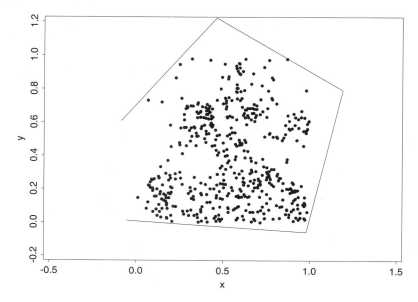

FIGURE 6.2. Scatter plot of the maple tree locations in the Lansing Woods with interactive boundary.

1. The *intensity* (number of points per unit area) of the point pattern does not vary over the bounded region, A. Formally, the number of points in planar region A with area $|A|$ follows a homogeneous Poisson distribution with mean $\lambda|A|$, where λ is the constant intensity.

2. There are no *interactions* among the points—points neither inhibit nor encourage each other. Formally, given n points with locations denoted by the vectors $\mathbf{x}$ in a region A, the $\mathbf{x}$ are an independent random sample from the uniform distribution on A.

We will look at techniques for examining the validity of this hypothesis using S+SPATIALSTATS in the following sections.

6.2.1   Visual Methods

Figure 6.3 shows two examples of non-random point patterns, both generated by the simulation methods detailed in section 6.4. The left plot shows apparent clustering, while the right plot indicates regularity.

Unlike these plots, the Lansing Woods data displayed in figure 6.1 is very dense, and no spatial pattern is immediately obvious. The Lansing Woods data is an example of a *bivariate* point pattern, with two types of points, hickory or maple trees. Plot these species separately as follows:

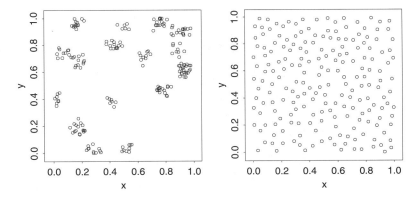

FIGURE 6.3. Example of two apparently non-random point patterns.

```
> hick.spp <- spp(lansing[lansing$species=="hickory",],
+     boundary=bbox(x=c(0,1), y=c(0,1)))
> maple.spp <- spp(lansing[lansing$species=="maple",],
+     boundary=bbox(x=c(0,1), y=c(0,1)))
> par(mfrow=c(1,2))
> par(pty="s")
> plot(hick.spp, main="Hickories")
> plot(maple.spp, main="Maples")
```

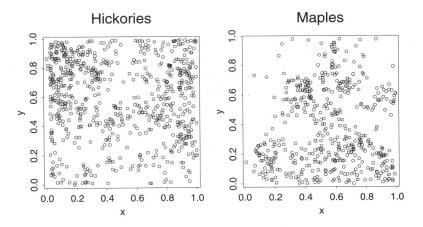

FIGURE 6.4. Scatter plots of the Lansing Woods tree locations for the two species.

We have created the hickory and maple point pattern objects with the unit square boundary common to the whole data set. The plots in figure 6.4 show that there may be interaction between the two tree species. It may be that the presence of one species inhibits the presence of the other species. If you have a color monitor, try overlaying one species with the other using the generic function **points** as follows (plot not shown):

```
> plot(hick.spp)
> points(maple.spp, col=3)
```

For a non-color monitor, you might use (plot not shown):

```
> points(maple.spp, pch="x")
```

Even if the whole set of tree locations proves to be spatially random, knowledge of species reveals an interesting spatial component to this data. Visual techniques may tell us all we need to know (as in the case of figure 6.3), but more formal tests of CSR are desirable.

6.2.2   Nearest Neighbor Methods

Nearest neighbor distances provide an objective method for looking at small scale interactions between points. These distances can be easily calculated in S+SPATIALSTATS, using the function **find.neighbor**. See section 5.1.3 for details on the use of **find.neighbor**. The nearest neighbor statistics discussed in the following sections can be used to compare your data to characteristics of a random pattern.

Point-to-Point Nearest Neighbor Statistic

Define a nearest neighbor distance d_i as the distance from the ith point to the nearest other point in A, the bounded region of interest. The empirical distribution function(EDF) of these *point-to-point nearest neighbor distances* can be used for comparison to a CSR process. The EDF is:

$$\widehat{G}(y) = n^{-1} \sum_{d_i \leq y} 1$$

where n is the number of points in A.

For the Lansing Woods data, calculate and plot $\widehat{G}$ using the function **Ghat** as follows:

```
> par(mfrow=c(1,2))
> hick.ghat <- Ghat(hick.spp)
> title(main="Hickories")
> maple.ghat <- Ghat(maple.spp)
> title(main="Maples")
```

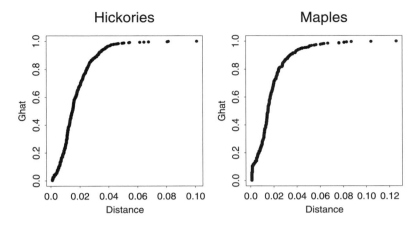

FIGURE 6.5. Plots of the EDF of point-to-point nearest neighbor distances for the two species in the Lansing Woods data: hickory (left), maple (right).

Each EDF is automatically plotted by `Ghat` unless argument `plot=F`. The distances y can be specified by the argument `dist.ghat`. If not specified, all nearest neighbor distances are used. The plot in figure 6.5 shows $\widehat{G}(y)$ for the two species. If there is clustering in the data, we would expect to see an excess of short distance neighbors. If there is regularity in the data, there will be an excess of long distance neighbors.

$\Longrightarrow$ **Hint:**   The function `Ghat` runs faster if used with the default distances.

Note: There are 703 locations for hickory trees in the `lansing` data frame, as in Diggle [1983]. The object `hick.ghat` created above reveals that two of these locations are identical by returning only 701 unique nearest neighbor distances.

The plot in figure 6.6 shows the $\widehat{G}$ plots for the two point patterns plotted in figure 6.3, for comparison to the Lansing Woods data. The left plot is an example of $\widehat{G}(y)$ for a point pattern exhibiting apparent clustering, and the right plot is an example involving apparent regularity.

Alternatives to visual judgement of the $\widehat{G}$ plot include testing against the theoretical distribution, or simulation techniques. The theoretical distribution function $G(y)$ for CSR is only available in closed form if edge effects are ignored. This case will be discussed in section 6.2.2.

When edge effects need to be considered, we can assess the hypothesis of CSR using Monte Carlo techniques. For example, we can simulate the EDF of nearest neighbor distances from s realizations of a CSR process on A, the region containing the original point pattern. The average of the s simulations provides a reference line, and the maximum and minimum provide a simulation *envelope*. If $\widehat{G}$ calculated from the data falls outside

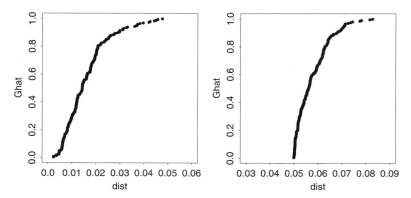

FIGURE 6.6. Plots of the EDF of point-to-point nearest neighbor distances for two apparently non-random point patterns.

this envelope either at short distances or at long distances, there is evidence against CSR.

For the Lansing Woods data, A is a square plot which has been scaled to unity. Thus, the estimated intensity (assumed to be constant for CSR) is simply equal to the total number of points. We can simulate a comparable Poisson (CSR) process using the S+SPATIALSTATS function make.pattern. Write a function to compute $\widehat{G}$ for s CSR patterns on a rectangle with n points as follows:

```
> ghat.env <- function(n, s, dist,
+       boundary=bbox(x=c(0,1),y=c(0,1))){
+ # n is the number of points to be simulated
+ # s is the number of simulations
+ # dist are the distances on which to calculate Ghat
+       hold <- matrix(0, s, length(dist))
+       for(i in 1:s) {
+            hold[i,  ] <- Ghat(make.pattern(n,
+                 boundary=boundary), dist.ghat=dist,
+                 plot=F)[,2]
+       }
+ # each row of hold contains the EDF for a single
+ # CSR simulation on the unit square
+       mn <- apply(hold,2,mean)
+       Up <- apply(hold,2,max)
+       Down <- apply(hold,2,min)
+       return(data.frame(mn, Up, Down))
+ }
```

The function `make.pattern` is discussed at length in section 6.4.

Create and plot simulation envelopes for the Lansing Woods maples $\widehat{G}$ statistic as follows:

```
> maple.env <- ghat.env(n=514, s=20,
+     dist=unique(maple.ghat[,1]))
```

We have computed the EDFs for 20 simulated CSR point patterns on the same distances as the Lansing Woods maples $\widehat{G}$. Notice that we have not used the argument `boundary`, since it defaults to the unit square.

If we use the mean of the 20 simulations as an estimate of the population value, plotting this versus the EDF for the data should approximate a straight line under CSR. Create this plot with minimum and maximum simulation envelopes as follows:

1. The distances used to calculate `maple.env` appear multiple times in `maple.ghat`. Create an index vector to plot an elongated `maple.env`, using the S-PLUS function `tapply`.

   ```
   > ind <- tapply(maple.ghat[,1], maple.ghat[,1])
   ```

2. Plot the estimate for the CSR $G(y)$ versus $\widehat{G}(y)$ for the maple data and for the envelopes.

   ```
   > par(mfrow=c(1,1))
   > plot(maple.env$mn[ind], maple.ghat[,2], type="line")
   > lines(maple.env$mn[ind], maple.env$Up[ind], lty=2)
   > lines(maple.env$mn[ind], maple.env$Down[ind], lty=2)
   ```

The plot in figure 6.7 shows strong evidence for clustering of maple trees. It is important to note that this evidence includes any effects from interaction with hickory trees. If we had restricted our simulations to the polygon containing just the maples, it is likely that this evidence would be weaker.

Origin-to-Point Nearest Neighbor Statistic

A test statistic related to $\widehat{G}$ is defined by overlaying a $k \times k$ grid on the region A, then comparing the distances from the m resulting origins to their nearest neighbors. Let e_i be the distance from the ith origin to the closest of the n points in the data. The empirical distribution function of these *origin-to-point nearest neighbor distances* is:

$$\widehat{F}(x) = m^{-1} \sum_{e_i \leq x} 1$$

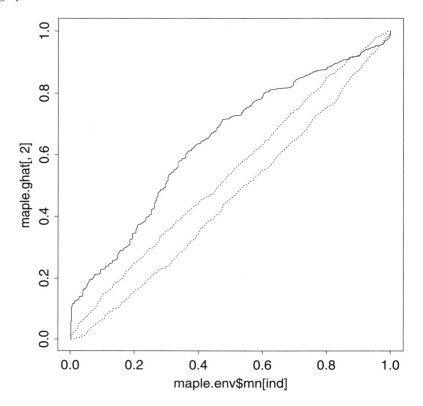

FIGURE 6.7. EDF plot of point-to-point nearest neighbor distances for the Lansing Woods maple data with upper and lower simulation envelopes from 20 simulations of a CSR process.

Find $\widehat{F}(x)$ for both species of the Lansing Woods data using the function Fhat as follows:

```
> par(mfrow=c(1,2))
> hick.fhat <- Fhat(hick.spp)
> title(main="Hickories")
> maple.fhat <- Fhat(maple.spp)
> title(main="Maples")
```

The $\widehat{F}$ statistics for the hickories and the maples are plotted in figure 6.8. The interpretation of the $\widehat{F}$ plot is opposite that of the $\widehat{G}$ plot. An excess of high distance values is interpreted as clustering. As before, we could compare this statistic to simulations from a CSR process for a visual interpretation. Instead we will assume there are no significant edge effects, and compare the maple $\widehat{F}$ to the theoretical CSR $F(x)$.

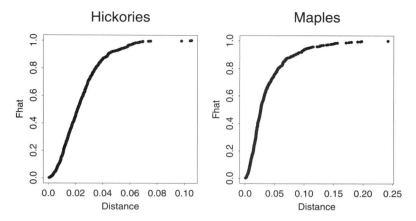

FIGURE 6.8. Plots of the EDF of origin-to-point nearest neighbor distances for the two species in the Lansing Woods data: hickory (left), and maple (right).

As discussed previously, there is a closed-form formula for the theoretical distribution functions $G(y)$ and $F(y)$ if edge effects can be ignored, and the number of points is relatively large. Edge effects are most likely to be a problem for small n or for elongated polygons. The theoretical distribution functions $G(y)$ and $F(y)$ are equal when CSR holds:

$$G(y) = F(y) = 1 - exp(-\pi\lambda y^2)$$

where λ is the constant intensity. For comparison to the EDF, substitute $\widehat{\lambda}$, the observed number of points per unit area. Then plot either nearest-neighbor statistic against this theoretical distribution function, or against each other, as a visual test of CSR. For the maple data:

```
> maple.edf <- 1 - exp(-514*pi*(maple.fhat[,1]**2))
> plot(maple.edf, maple.fhat[,2])
> abline(0,1)
```

The plot in figure 6.9 again shows the clustering of the Lansing Woods maple trees. In figure 6.10, we have created the same plots for the clustered and regular patterns plotted in figure 6.3 for comparison.

6.3 Examining First- and Second-Order Properties

In this section we begin looking at the construction of a model to describe a spatial point pattern. First we must determine whether the underlying

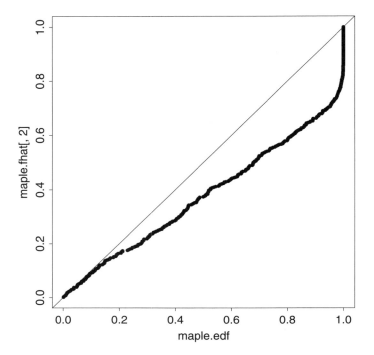

FIGURE 6.9. Plot of the theoretical EDF of origin-to-point nearest neighbor distances under CSR vs. the EDF for the Lansing Woods maples.

spatial process is *stationary* and *isotropic*. These terms have been defined in previous chapters. In this context, a spatial point process is:

- *stationary* if the intensity is constant, and the *second-order intensity* depends only on the direction and distance between pairs of points (and not on their absolute locations).

- *isotropic* if it is stationary, and the second-order intensity depends only on the distance between pairs of points, and not on the direction.

The *second-order intensity* of a spatial point process is a measure of the spatial dependence between points (Gatrell et al., 1995). The following sections show how to investigate these properties using S+SPATIALSTATS.

6.3.1 Intensity

First-order properties of a spatial point process describe how the mean number of points per unit area (the intensity) varies through space. For

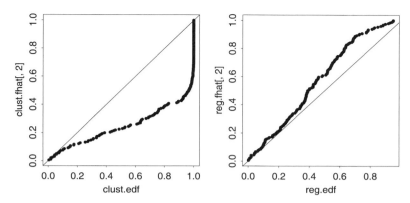

FIGURE 6.10. Plots of the theoretical EDF of origin-to-point nearest neighbor distances under CSR vs. the EDF for a clustered process (left), and a regular process (right).

a stationary process, the intensity λ is assumed to be constant over the bounded region of interest A.

To estimate the intensity of a spatial point pattern in S+SPATIALSTATS we use the function intensity. There are four methods available: *basic, binning, kernel, and gauss2d*. The argument method="basic" results in a single point estimate of the intensity, the number of points divided by the area of the bounded region. The other three methods estimate the intensity locally over the total region A, and return a list containing smoothed intensity estimates which may vary over A. The image function can be used to visualize this variation and to assess the hypothesis of stationarity for the point pattern.

The *binning* method uses a two-dimensional histogram to form rectangular bins. The counts in these bins are smoothed using a loess smoothing algorithm. Use the binning method for the maple data as follows:

```
> maple.bing <- intensity(maple.spp, method="binning",
+       nx=30, ny=30, span=.1)
```

The arguments nx and ny control the number of rectangles used. If not specified, they default to the square root of the total number of points in the pattern. The argument span is used by loess, the smoothing function (see the loess help file for possible and recommended values for span).

The function intensity returns a list containing x, y, and z values suitable for plotting using the S-PLUS function image, as follows:

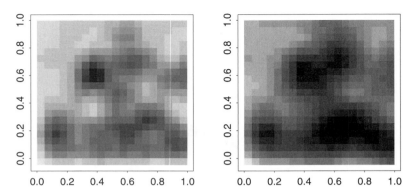

FIGURE 6.11. Image plots of the intensity for the Lansing Woods maples, estimated using the binning method with span = .1 (left), and span = .2 (right).

```
> image(maple.bing)
> image(intensity(maple.spp, method="binning", nx=30,
+      ny=30, span=.2))
```

Both grayscale plots are displayed in figure 6.11. The right plot shows that the larger span, 0.2, results in a smoother map (or try nx=50 and ny=50).

If method="kernel", intensity estimates the intensity at each point of the pattern, using a weighted function of the points in the surrounding *region of influence*. The *kernel* method uses a quartic kernel estimator to estimate the intensity of a spatial point pattern, using a bandwidth—the radius of the region of influence—given by the argument bw. Use the quartic kernel estimator on the maple data as follows:

```
> maple.kern <- intensity(maple.spp,method="kernel",bw=1)
> image(maple.kern)
```

The argument bw can differ in the x and y directions, that is, a vector with 2 values can be input. The user is required to provide a value for this argument for methods *kernel* and *gauss2d*. The resulting grayscale image is the left plot in figure 6.12.

The *gauss2d* method uses a Gaussian kernel estimate of the intensity. Use this method on the maple data, using the same bandwidth as above, as follows:

```
> maple.gaus <- intensity(maple.spp,method="gauss2d",bw=1)
> image(maple.gaus)
```

The resulting grayscale plot is the right-hand plot in figure 6.12.

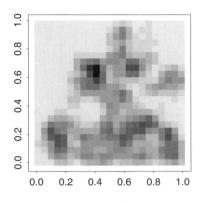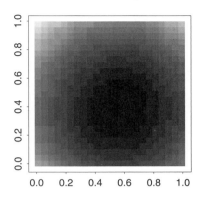

FIGURE 6.12. Image plots of the intensity for the Lansing Woods maples, estimated using the quartic kernel function (left), and the Gaussian kernel function (right).

To create a three-dimensional plot of the intensity of a spatial point pattern in S+SPATIALSTATS, use the Trellis graphics function `wireframe` as follows:

```
> trellis.device(color=F)
> mgrid <- expand.grid(x=maple.bing$x, y=maple.bing$y)
> mdf <- data.frame(x=mgrid$x, y=mgrid$y,
+      z=c(maple.bing$z))
> wireframe(z~x*y, data=mdf, drape=T)
```

We have used `trellis.device` with `color=F` for a black and white rendition. The result is plotted in figure 6.13.

All of the intensity estimation and visualization techniques used in this section show that the intensity of the maple trees in the Lansing Woods appears to vary more than would be expected by random fluctuations. This might be due to the deficit of maple trees in the north corners of the plot, which might be explained by interaction with hickory trees.

6.3.2 The K-function

The second-order properties of spatial point processes describe how the interaction or spatial dependence between points varies through space. These properties are usually described by the second-order intensity of the spatial point pattern. An alternative description of the second order properties is defined by the *K-function*:

$$K(d) = \lambda^{-1} E[number\ of\ points \leq distance\ d\ of\ an\ arbitrary\ point].$$

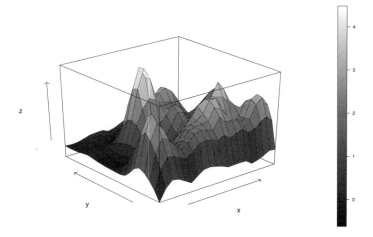

FIGURE 6.13. Wireframe plot of the intensity for the Lansing Woods maples, estimated using the binning method.

where λ is the intensity, and $E[\ ]$ denotes the expectation. The advantage of the K-function is that the theoretical value for $K(d)$ is known for several useful models of spatial point processes. For example, the K-function for a homogeneous process with no spatial dependence is πd^2. If there is clustering, we would expect an excess of points at short distances, leading to $K(d) > \pi d^2$ for small d's. Similarly, we would expect $K(d) < \pi d^2$ for a regularly spaced pattern.

Ripley's [1976] estimator for the above K-function is:

$$\widehat{K}(d) = n^{-2}|A| \sum \sum_{i \neq j} w_{i,j}^{-1} I_d(d_{i,j}).$$

where n is the number of points in region A with area $|A|$, $d_{i,j}$ is the distance between the ith and jth points, $w_{i,j}$ is the proportion of the circle with center at i and passing through j which lies within A, and $I_d(d_{i,j})$ is an indicator function which is 1 if $d_{i,j} \leq d$. This estimate includes an adjustment for edge effects.

Calculate Ripley's K-function for the Lansing Woods data using the S+SPATIALSTATS function Khat as follows:

```
> lans.khat <- Khat(lansing.spp)
> lans.d <- lans.khat$values[,"dist"]
> lines(lans.d, pi*lans.d**2)
> maple.khat <- Khat(maple.spp)
> maple.d <- maple.khat$values[,"dist"]
```

```
> lines(maple.d, pi*maple.d**2)
> hick.khat <- Khat(hick.spp, plot=F)
```

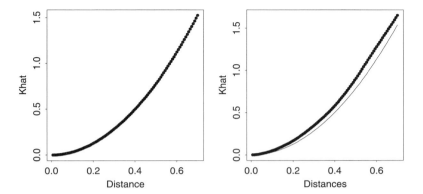

FIGURE 6.14. Plots of the Lansing Woods K-function estimates for all trees (left) and for maples only (right). Added lines are for a CSR process.

The function Khat calculates and plots $\widehat{K}(d)$, the estimate of $K(d)$ defined above. We have used the argument plot=F for the hickory trees. The matrix values returned by Khat contains a column with distances, and the corresponding values for $\widehat{K}$. The two plots created by calls to Khat above are shown in figure 6.14. We have added the reference lines which correspond to spatial randomness. The left plot shows that the K-function for the combined Lansing Woods data is very close to what would be expected for a CSR process. The right plot indicates that $K(d) > \pi d^2$ for the maple trees, or that clustering occurs.

There is also a function available in S+SPATIALSTATS for calculating $\widehat{L}(d)$ for $L(d) = \sqrt{K(d)/\pi}$, a variance stabilizing transformation. $L(d)$ is a straight line for a homogeneous Poisson random process. Plot $\widehat{L}(d)$ for the hickory data as follows:

```
> par(mfrow=c(1,1))
> hick.lhat <- Lhat(hick.spp)
> abline(0,1)
```

The function abline adds a line with intercept 0 and slope 1 to the current plot. The plot in figure 6.15 shows clustering in the hickory data at small distances.

We can evaluate the significance of the clustering by adding simulation envelopes and averages to the plot as we did with $\widehat{G}$ in section 6.2.2. Add envelopes to both species $\widehat{K}$ plots using the function Kenv as follows:

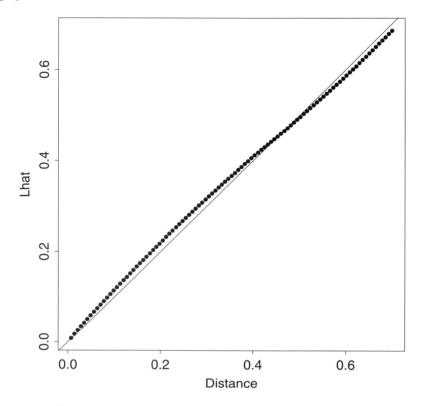

FIGURE 6.15. Plot of the Lansing Woods L(d) estimates and that of a CSR process for the hickory trees.

1. Plot `maple.khat`, restricting the range to distances less than 0.25 for clarity:

```
> par(mfrow=c(1,2))
> plot(maple.khat$values
+      [maple.khat$values[,1] < 0.25,])
```

2. Calculate and plot the simulation envelopes for the maples:

```
> maple.kenv <- Kenv(maple.spp, nsims=25, add=F)
> ind <- (maple.kenv$dist < 0.25)
> lines(maple.kenv$dist[ind], maple.kenv$lower[ind])
> lines(maple.kenv$dist[ind], maple.kenv$upper[ind])
```

3. Follow the same procedures for the hickories:

```
> plot(hick.khat$values[hick.khat$values[,1] < 0.25,])
> hick.kenv <- Kenv(hick.spp, nsims=25, add=F)
```

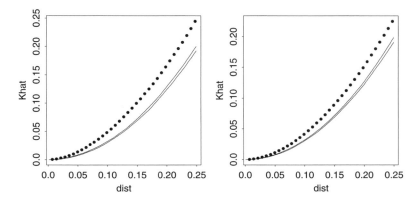

FIGURE 6.16. Plots of the Lansing Woods K-function estimates for the two species with simulation envelopes added.

```
> ind2 <- (hick.kenv$dist < 0.25)
> lines(hick.kenv$dist[ind2], hick.kenv$lower[ind2])
> lines(hick.kenv$dist[ind2], hick.kenv$upper[ind2])
```

The plots in figure 6.16 show evidence of clustering for both species.

6.4   Simulating Point Patterns

Many analysis procedures for spatial point patterns are based on Monte Carlo methods. In S+SPATIALSTATS, the function make.pattern generates a random two-dimensional set of points following one of five basic models (*Poisson, binomial, SSI, Strauss, and cluster*). For example, to generate and plot a CSR process on the unit square:

```
> par(mfrow=c(1,1))
> plot(make.pattern(n=100))
```

The process="binomial" option is the default for make.pattern, so the plot in figure 6.17 is a binomial process (CSR) process with 100 points. This is equivalent to a homogeneous Poisson process conditioned on a fixed number of points. A CSR process can also be simulated by specifying process="poisson". In this case, the argument lambda is given instead of n, and the intensity, rather than the exact number of points is the focus. These patterns can be used to create simulation envelopes around the statistics discussed earlier in this chapter.

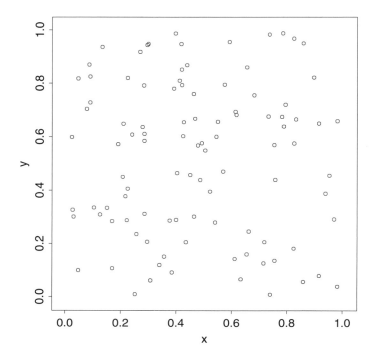

FIGURE 6.17. Plot of 100 points from a completely random (binomial) spatial process generated by the function `make.pattern`.

If you have seen evidence of regularity in your data, try generating an SSI or a Strauss process to model your data. The SSI (Sequential Spatial Inhibition) process keeps only those pairs of a random process which are more than a given *radius of inhibition* from each other. Generate and plot 100 points for an SSI process with a radius of inhibition equal to 0.1 as follows:

```
> par(mfrow=c(1,2))
> plot(make.pattern(100, process="SSI", radius=.1,
+       boundary=bbox(x=c(0,1),y=c(0,2))))
```

The left plot in figure 6.18 displays the SSI pattern generated. We have specified a 1×2 rectangular bounded region in our call to `make.pattern`.

The `"Strauss"` process (Ripley, 1981, p. 166) generates a point pattern which allows only a specified fraction of points to remain within the radius of inhibition. This fraction is specified by the argument `cpar`, which defaults to 0. The generation proceeds as follows:

1. A random point is placed in A, the bounded region.

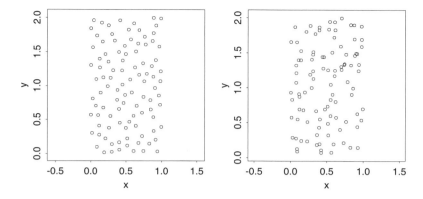

FIGURE 6.18. Plot of an SSI spatial process (left) and a Strauss spatial process (right), both generated by the function `make.pattern`.

2. Subsequent random points are generated, and are kept with probability c^s, where c is the inhibition parameter on [0,1], and s is the number of existing points closer than r to the potential new point.

Generate and plot a Strauss point pattern with radius of inhibition of 0.1 and inhibition parameter of .5 as follows:

```
> plot(make.pattern(100, process="Strauss",radius=.1,
+       cpar=.5, boundary=bbox(x=c(0,1),y=c(0,2))))
```

The right-hand plot in figure 6.18 displays the result.

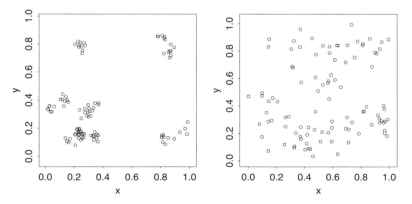

FIGURE 6.19. Plots of two Poisson cluster processes, both generated by the function `make.pattern`.

Warning: If the radius of inhibition is too large, it may be impossible or nearly impossible to generate the requested number of points.

The `process="cluster"` option produces a Poisson parent-daughter process, with the argument `cpar` used for the mean of the parent process, and the argument `radius` used for the radius of the clusters. Generate Poisson cluster processes on the unit square as follows:

```
> plot(make.pattern(100, process="cluster",
+      radius=.05, cpar=15))
> plot(make.pattern(100, process="cluster",
+      radius=.15, cpar=15))
```

Both plots in figure 6.19 have a parent Poisson process with mean 15. The left-hand plot has a smaller radius for the daughter process, resulting in more distinct clusters.

7

S+SPATIALSTATS and GIS

This chapter introduces the Geographic Information System (GIS) user to functions in S-PLUS and S+SPATIALSTATS that may be used for the analysis of spatial data from a GIS database. The main features of S+SPATIALSTATS are illustrated through specific examples using data imported from the ARC/INFO GIS. The import is done with S+GISLINK, a module for S-PLUS that allows data transfer between S-PLUS and ARC/INFO. The examples in this chapter are meant to supplement the non-GIS examples used in chapters 4–6, by showing how spatial data analysis in S+SPATIALSTATS might be implemented using several types of data from GIS.

In this chapter you will learn about the following:

- GIS data, especially ARC/INFO coverages and spatial data analysis (section 7.1).

- Spatial data analysis of a point coverage (section 7.2).

- Spatial data analysis of a polygon coverage (section 7.3).

- Spatial data analysis of a grid (section 7.4).

7.1   ARC/INFO Coverages and Spatial Data Analysis

In chapter 1 we described the types of spatial data that S+SPATIALSTATS was designed for, and in chapters 4 through 6 we demonstrated some of the analytical techniques that are available for use with spatial data. By the very nature of the tool, ARC/INFO is a depository for an abundance of spatially referenced data for which spatial data analyses are appropriate. Table 7.1 shows the possible correspondences between ARC/INFO coverage types and these spatial data types, along with related spatial analysis methods. Corresponding S+SPATIALSTATS functions are also listed.

TABLE 7.1. Key to S+SPATIALSTATS Functions.

Cov	Spatial Data	Analysis	Functions

point	non-gridded geostatistical data	Variogram estimation
		
		Variogram modeling
		Kriging
			`variogram`
			`variogram.cloud`
			`model.variogram`
			`krige`
			`predict.krige`

point	point pattern	Nearest neighbor
		→ G statistic
		→ F statistic
		Ripley's K functions
			
			`Ghat`
			`Fhat`
			`Khat, Kenv`
			`Lhat, Lenv`

polygon	irregular lattice	Spatial autocorrelation
		→ Geary
		→ Moran
		Spatial regression
			`spatial.neighbor`
			`find.neighbor`
			`spatial.cor`
			`slm`

grid	gridded geostatistical data	Variogram estimation
		
		Variogram modeling
		Kriging
			`variogram`
			`variogram.cloud`
			`model.variogram`
			`krige`
			`predict.krige`

grid	regular lattice	Spatial autocorrelation
		→ Geary
		→ Moran
		Spatial regression
			`spatial.neighbor`
			`find.neighbor`
			`spatial.cor`
			`slm`

7.1.1 ARC/INFO Coverage Types and Spatial Data Types

In ARC/INFO there are three basic coverage types; namely, point, polygon, and line (arc) coverages. One can also have grids which are used for most of the analytical functions in ARC/INFO. In S+SPATIALSTATS there are tools to analyze point and polygon coverages, as well as grids.

For example, some point coverages may be analyzed as non-gridded geostatistical data when the points represent fixed locations at which observations of a random variable were taken or measured. Typical analyses include variogram estimation for the modeling of spatial correlation and kriging for making predictions that incorporate the modeled correlation. Other point coverages may be better analyzed as spatial point patterns; in this case, the point locations themselves are considered random. Typical analyses for spatial point patterns include calculating statistics that help determine the nature of the pattern—are the locations clustered, dispersed, or randomly located?

Polygon coverages in ARC/INFO are particularly suited to treatment as irregular lattice data. The polygons define regions with the preservation of neighbor information. The collection of polygons forms an indexed set of regions for which tests of spatial autocorrelation or the use of spatial linear models are appropriate.

Grids can be analyzed in S+SPATIALSTATS as either gridded geostatistical data, or as regular lattice data. Again, geostatistical data is defined for fixed locations, and lattices are defined for collections of spatial regions. Lattice data typically have some sort of associated neighborhood information. For gridded data, the distinction of corresponding data type may not be easy. One might choose analyses from either category of data, especially for exploratory data analysis.

In sections 7.2 through 7.4 we demonstrate some typical analyses that might be used for several types of data.

7.1.2 Spatial Referencing of Data for Use with S+SPATIALSTATS

There are several considerations for the ARC/INFO user prior to importing data into S-PLUS for spatial analysis. These issues, all related to the spatial referencing of data, are discussed below.

In order to answer questions about spatial relationships, locational information must be attached to a coverage before transfer to S-PLUS. While ARC/INFO maintains an internal representation of the spatial structure of coverage data, many functions in S-PLUS and S+SPATIALSTATS require explicit location information.

For point or polygon coverages, the ARC command ADDXY can be used to add the x, y coordinates of the label points to the .PAT file. For point coverages this is sufficient, as the locations of the label points represent the points themselves. In the case of polygon coverages, however, the coordinates of the label points can be arbitrary, having been defined during digitizing or through an ARCEDIT session.

Alternatively, polygon label points can be defined using the ARCEDIT command addlabels, resulting in x, y coordinates corresponding to the geometric centroids of the polygons. For spatial analyses that are based on distance measures, the use of geometric centroids is generally desirable. Thus, the extent of preprocessing that may be necessary before importing a polygon coverage into S-PLUS depends on the history of the coverage and the types of analyses to be done. Polygon label points may need adjusting prior to importing data into S-PLUS.

Calculations in S+SPATIALSTATS are based on a cartesian coordinate system. Users of ARC/INFO are provided with many options for creating map projections. As with analyses conducted in ARC/INFO, the user should be aware of the projections of the coverages, and should ensure that the locational coordinates are the same for any comparisons made across coverages.

7.2   Example: Analysis of a Point Coverage

Point coverages in ARC/INFO will generally correspond to spatial data that can be treated as non-gridded geostatistical data or as a spatial point pattern. The example we analyze in this section will be treated as geostatistical data. The techniques presented here are detailed in chapter 4. If you want to treat the data as a point pattern, chapter 6 contains details on the relevant tools. The point data represent fixed sampling locations at which rainfall measurements were taken.

7.2.1   Rainfall Coverage

The ARC/INFO **rainfall** coverage consists of 10-year monthly mean rainfall measurements. The measurements were taken at 565 locations irregularly located over an area of 20 degrees longitude and 14 degrees latitude, in the Amazon River Basin. The data was kindly provided by the EOS-Amazon Modeling Project at the University of Washington, a NASA supported interdisciplinary science team, in support of research and development. The monthly data represents modeled output from original ground stations supplied by the Departamento Nacional de Aguas e Energia Elétrica (DNAEE) in Brazil.

Note: This data set is not distributed as part of S+SPATIALSTATS. Exploratory maps of the **rainfall** coverage, generated in ARCPLOT using the POINTSPOT command, suggest both spatial and temporal patterns. Questions that one may want to ask include: 1) is the pattern of rainfall uniform throughout space; 2) do the rainfall data exhibit spatial correlation; and 3) is the spatial correlation the same for the wet and dry seasons. Answers can be explored by moving your data to S-PLUS and using some of the analytical tools available in S-PLUS and S+SPATIALSTATS, many of which have been described earlier in this manual.

7.2.2 Importing Rainfall Data into S-PLUS

The **rainfall** coverage can be copied to S-PLUS using the S+GISLINK `arcsplus` command with the `point` option. In ARC/INFO enter the following commands to export the **rainfall** coverage as an S-PLUS data frame.

```
Arc: copy rainfall rain-xy
Arc: addxy rain-xy
Arc: arcsplus rain-xy point rain.df
```

The ARC `COPY` command is used to make a working copy of the **rainfall** coverage. This step is *not* necessary, but will preserve the original coverage from direct modifications. The ARC `ADDXY` command adds the x, y coordinates of each point location to the `RAIN-XY.PAT` attribute table. This step *is* necessary to perform spatial analyses in S-PLUS.

The S+GISLINK command `arcsplus` copies data in the `RAIN-XY.PAT` table to the S-PLUS data frame `rain.df`. To view the items of the rainfall data frame, use the S+GISLINK `splusexecute` command with the S-PLUS function `names` as follows:

```
Arc: splusexecute 'names(rain.df)'
 [1] "AREA"        "PERIMETER"  "RAIN.XY."    "RAIN.XY.ID"
 [5] "MEAN1"       "MEAN2"      "MEAN3"       "MEAN4"
 [9] "MEAN5"       "MEAN6"      "MEAN7"       "MEAN8"
[13] "MEAN9"       "MEAN10"     "MEAN11"      "MEAN12"
[17] "X.COORD"     "Y.COORD"
```

These variables correspond to those obtained by a call to the ARC ITEMS command. The `rain.df` data frame has 18 data items. Of these data, the **AREA** and **PERIMETER** items will not be used for analysis since their values are zero for all points (a defining attribute of a point coverage). Further, the `RAIN.XY.` and `RAIN.XY.ID` items contain unique, sequential identifiers for each point and will not contribute to data analysis. Verify the contents of the first four items, and then create a working data frame using the remaining 14 items:

1. Set up an abbreviation for the `splusexecute` command to eliminate unnecessary typing.

   ```
   Arc: &sv se = splusexecute
   Arc: &abbrev &on
   ```

2. Check the values of `AREA`, `PERIMETER`, `RAIN.XY.`, and `RAIN.XY.ID` by looking at their range and number of records.

   ```
   Arc: se 'range(rainpat.frame$AREA)'
   [1] 0 0
   Arc: se 'range(rainpat.frame$PERIMETER)'
   [1] 0 0
   Arc: se 'length(rainpat.frame$RAIN.XY.)'
   [1] 565
   Arc: se 'length(rainpat.frame$RAIN.XY.ID)'
   [1] 565
   ```

3. Delete the four noninformative items from `rain.df`. Items are columns in the data frame, so we remove the corresponding 4 columns by name.

   ```
   Arc: se 'rain.no <- c("AREA","PERIMETER","RAIN.XY.",
              "RAIN.XY.ID")'
   Arc: se 'rain.keep <- setdiff(names(rain.df),
                        rain.no)'
   Arc: se 'rain.df <- rain.df[,rain.keep]'

   S-PLUS processing...
   Arc: se 'names(rain.df)'
    [1] "MEAN1"  "MEAN2"  "MEAN3"  "MEAN4"  "MEAN5"
    [6] "MEAN6"  "MEAN7"  "MEAN8"  "MEAN9"  "MEAN10"
   [11]"MEAN11" "MEAN12" "X.COORD" "Y.COORD"
   ```

4. Compute and add a new column for annual mean rainfall, for possible additional comparisons.

   ```
   Arc: se 'rain.df[,"annual"] <- apply(rain.df, 1,
                           FUN=function(x)
                           sum(x[1:12]))'
   ```

The `rain.df` data frame now has 14 items from the original **rainfall** coverage. In addition, there is a new column for annual rainfall. We are now ready to proceed with analysis of the rainfall data.

Note: The exploratory data analysis and spatial modeling that follows will be done exclusively using functions from S-PLUS and S+SPATIAL-STATS. For ease of use, we recommend that the user either invoke an S-PLUS session from within ARC/INFO by entering `Splus` at the **Arc:** prompt or quit ARC/INFO and invoke S-PLUS as a separate process.

7.2.3 Exploratory Data Analysis

In addition to the x,y locations, the `rain.df` data frame has 12 columns for the monthly mean rainfall measurements and 1 column for annual mean rainfall. Descriptive statistics for monthly mean rainfall can be obtained using `summary` as follows:

```
> summary(rain.df[,1:12])
      MEAN1             MEAN2              MEAN3
 Min.   : 27.8    Min.   : 22.63    Min.   : 47.53
 1st Qu.:251.6    1st Qu.:253.50    1st Qu.:266.10
 Median :281.6    Median :286.10    Median :296.60
 Mean   :267.8    Mean   :273.70    Mean   :291.10
 3rd Qu.:309.4    3rd Qu.:313.00    3rd Qu.:325.60
 Max.   :403.3    Max.   :505.70    Max.   :563.40
 .
 .
 .

      MEAN10            MEAN11             MEAN12
 Min.   : 28.98   Min.   : 46.45    Min.   : 42.7
 1st Qu.: 95.97   1st Qu.:113.20    1st Qu.:178.0
 Median :160.50   Median :198.00    Median :246.7
 Mean   :146.00   Mean   :174.70    Mean   :229.4
 3rd Qu.:186.10   3rd Qu.:227.60    3rd Qu.:288.4
 Max.   :248.10   Max.   :286.00    Max.   :370.6
```

The `summary` function returns the minimum, maximum, mean, median, and 1st and 3rd quartiles for each month (only 6 months shown). The summary statistics reveal differences in both the range and distribution of rainfall over the annual record.

To explore the data graphically first create a second data frame with one column containing all the rainfall values and a second column containing a factor variable identifying the month for each element in the first column.

This will allow conditioning on months in Trellis graphics displays. Use the `make.groups` function to create this data frame:

```
> # Attach the original data frame to save typing
> attach(rain.df)
> rain.df2 <- make.groups(
+       Jan=MEAN1, Feb=MEAN2, Mar=MEAN3, Apr=MEAN4,
+       May=MEAN5, Jun=MEAN6, Jul=MEAN7, Aug=MEAN8,
+       Sep=MEAN9, Oct=MEAN10, Nov=MEAN11, Dec=MEAN12)
> names(rain.df2) <- c("Rain","Month")
> # Make Month an ordered factor
> rain.df2$Month <- ordered(rain.df2$Month,
+       labels=month.abb)
> # Add the locations to the data frame for later plotting
> rain.df2$Easting <- rep(X.COORD, 12)
> rain.df2$Northing <- rep(Y.COORD, 12)
> detach("rain.df")
```

To look at the distributions of the monthly averages use the S-PLUS `histogram` function:

```
> histogram(~ Rain | Month, data=rain.df2,
+       nint=25,  as.table=T)
```

The resulting histograms are shown in figure 7.1. A variety of skewed and bimodal distributions are displayed; spatial uniformity is doubtful.

Note: The `as.table=T` argument to `histogram` lays out the panels of the plot with Jan, the first level of the ordered factor `Month`, at the top left. This argument is not available in S-PLUS 3.3. If the argument is not used the Jan panel will appear at the lower left.

Begin exploring the spatial structure of the rainfall data by looking at scatterplots of mean monthly rainfall versus latitude and versus longitude. Create the scatterplots shown in figure 7.2 as follows:

```
> xyplot(Rain ~ Northing | Month, data=rain.df2, as.table=T,
+       pch='.',  panel=function(x,y,...) {
+               panel.xyplot(x,y,...)
+               panel.loess(x,y,degree=2,...)
+       }
+ )
```

The lines drawn by the `panel.loess` are used to highlight the forms of the trends in the data. Figure 7.2 reveals two clear trends in monthly mean rainfall with changes in latitude (`Northing`): the groups December

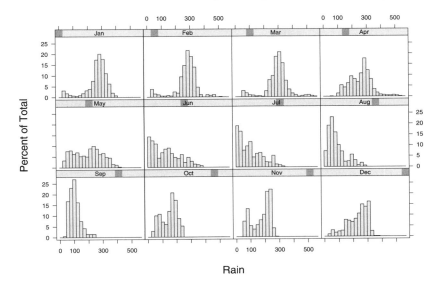

FIGURE 7.1. Histograms of monthly mean rainfall.

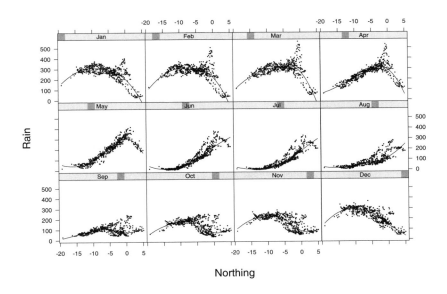

FIGURE 7.2. Scatterplots of monthly mean rainfall versus latitude.

through March and June through August have similar within-group trends. The months of April, May, September, and October seem to be transitional between the two trends. Scatterplots as a function of `Easting` are similarly

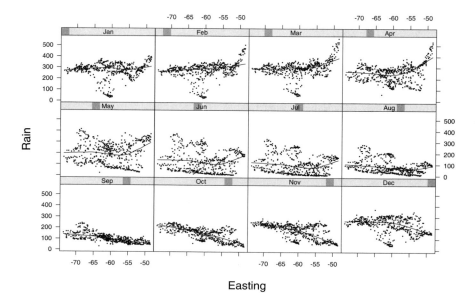

FIGURE 7.3. Scatterplots of monthly mean rainfall versus longitude.

created and are shown in figure 7.3. Specific trends along longitude are less striking, though some interesting patterns arise.

7.2.4 Spatial Modeling

The histograms and scatterplots indicate that rainfall pattern is governed by both temporal and spatial processes. As such, the analysis might proceed by first modeling the spatial component of the data for individual time periods—this assumes independence of the spatial and temporal effects. We proceed by looking at some methods in S+SPATIALSTATS for characterizing the spatial structure of the data.

Local Regression Model

In order to incorporate the apparent spatial structure of the rainfall data into a model, we can use local regression methods to model rainfall as a function of the x, y locations. Local regression (loess) models use nonparametric locally weighted regression to model a surface, given the predictors. The loess model is fitted as a smooth function of all the predictors, including interactions. A loess spatial model of the monthly mean rainfall for each month was created using the S-PLUS functions `loess` and `predict.loess`. See section 3.2.2 for details on the use of the these functions. For displaying the loess surface, predictions are made over a grid restricted to the convex hull of the spatial locations. The grid is created using the functions `chull` and `poly.grid`. See the examples in section 4.3.

```
> chull.XY <- chull(rain.df$X.COORD, rain.df$Y.COORD)
> chull.XY <- cbind(rain.df$X.COORD[chull.XY],
+         rain.df$Y.COORD[chull.XY])
> grid.rain <- poly.grid(chull.XY, nx = 30, ny = 30)
> grid.rain <- data.frame(grid.XY)
> names(grid.rain) <- names(rain.df2)[3:4]
> Resid.lo <- NULL
> Pred.lo <- NULL
# Loop over each month, use the subset argument with loess
# to fit each month separately:
> for(i in levels(rain.df2$Month)) {
+         lobj <- loess(Rain ~ Easting * Northing,
+                 data = rain.df2,
+                 subset = (rain.df2$Month == i),
+                 span = 0.5, degree = 2, normalize = F)
+         Resid.lo <- c(Resid.lo, resid(lobj))
+         Pred.lo <- c(Pred.lo, predict(lobj, grid.rain))
+ }
# Add the loess residuals to the rain.df2 data frame
> rain.df2$Resid.lo <- Resid.lo
# Create a new data frame with the predictions,
# the grid locations and an ordered Month factor:
> rain.predict.df <- data.frame(Pred.lo=Pred.lo,
+         Month=ordered(rep(month.abb,
+                 rep(dim(grid.rain)[1], 12)),
+                 levels=month.abb),
+         Easting=rep(grid.rain$Easting, 12),
+         Northing=rep(grid.rain$Northing, 12))
```

The Trellis graphics functions `levelplot` and `wireframe` can be used to obtain 2-D and 3-D plots of the predicted surfaces, respectively. The plots of the predicted surfaces, shown in figure 7.4 and figure 7.5 were created as follows:

```
> levelplot(pred.loess ~ Easting * Northing | Month,
+         data=rain.predict.df, contour=T, pretty=T,
+         as.table=T)
> wireframe(pred.loess ~ Easting * Northing | Month,
+         data=rain.predict.df, zlab="fit", as.table=T)
```

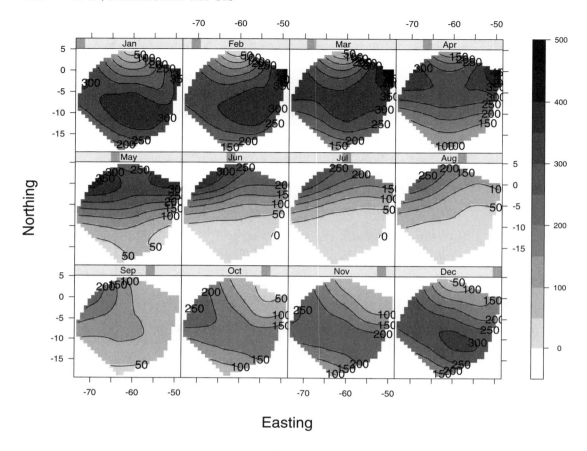

FIGURE 7.4. Local regression model using spatial locations as predictors—levelplots (filled contour plots) of the fitted surface for monthly mean rainfall.

Each of the panels in the displays have the same scaling so comparisons can be made between months. The plots show both spatial and temporal trends. There is a strong spatial trend in the wet months (Jan, Feb, Mar, Apr) with high values in the central basin and the northeast. In the dry season the highest rainfall is in the northwest. The middle of the southern area of the basin always has the lowest rainfall each month.

Variogram Modeling

The loess model assumes that the observations are independent. Data such as rainfall, however, are often governed by an underlying small-scale stochastic process in addition to any large-scale trends. This small- to micro-scale variation may be modeled as spatial autocorrelation and incor-

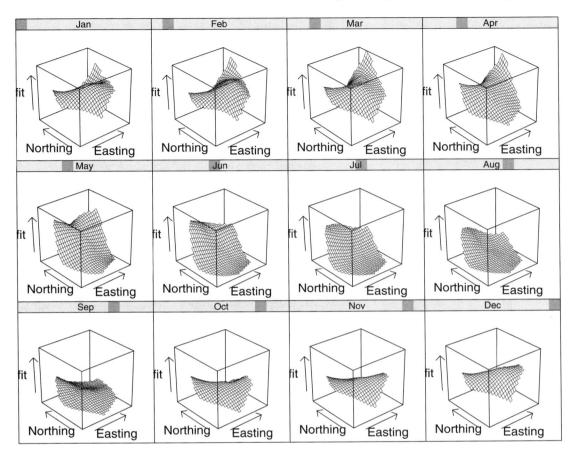

FIGURE 7.5. Local regression model using spatial locations as predictors—wireframe plots of the fitted surface for monthly mean rainfall.

porated into estimation procedures. Construction of variograms is one way to explore and model spatial correlation for continuous data. See section 4.1 for a complete discussion on variogram estimation using S+SPATIALSTATS.

We explore the spatial correlation structure of the rainfall data by constructing empirical variograms using the original data. Directional variograms are better than omnidirectional variograms in this case, since trends have been observed. Compute the variograms for each month and combine the results in a single data frame:

```
> vg.df <- NULL
> mon.vec <- NULL
> for(i in levels(rain.df2$Month)) {
```

```
+        vgi <- variogram(Rain ~ Easting + Northing,
+            data=rain.df2,
+            subset=(rain.df2$Month == i),
+            azimuth=c(0,45,90,135), tol.azimuth=22.5)
+        vg.df <- rbind(vg.df, vgi[, c("distance",
+            "gamma", "azimuth")])
+        mon.vec <- c(mon.vec, rep(i, dim(vgi)[1]))
+ }
# Add mon.vec to vg.df data frame as an ordered factor:
> vg.df$Month <- ordered(mon.vec, levels=month.abb)
```

The variograms displayed in figure 7.6 were plotted with:

```
> xyplot(gamma ~ distance | Month*azimuth,
+     data=vg.df, as.table=T, pch='.',
+     layout=c(12,4,1))
```

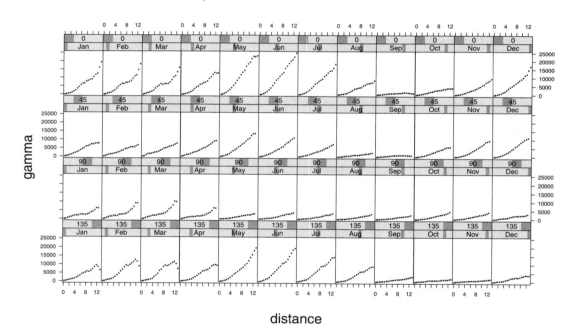

FIGURE 7.6. Empirical directional variograms for each month's rainfall for 0, 45, 90, and 135 degrees clockwise from north.

The directional variograms are almost all increasing and essentially unbounded. This is likely due to the previously observed trends. To verify this, compare these variograms with directional variograms fit to the loess model residuals. (These residuals were saved as Resid.lo in the rain.df2 data frame.) The resulting variograms, shown in figure 7.7, are essentially bounded and reveal some short range spatial correlation.

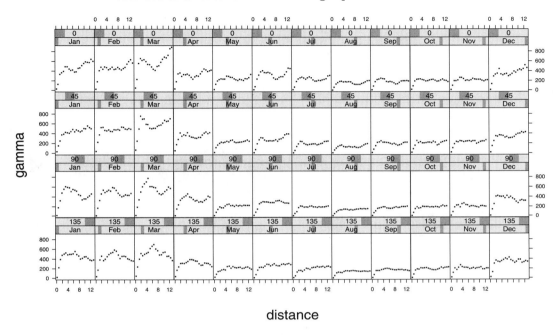

FIGURE 7.7. Empirical directional variograms for each month's rainfall based on the residuals from a loess model.

The directional variograms of the residuals for any month are very similar across directions. At this point, we will assume that the spatial correlation structures for each month can be adequately modeled as isotropic processes. See section 4.1.4 for methods used to identify and correct for anisotropy.

Theoretical models can be fitted to the empirical variograms using the interactive S+SPATIALSTATS function `model.variogram`. See sections 4.2.1 and 4.2.2 for details on theoretical variogram models and model fitting. A spherical model was selected and fit individually to the loess residuals for each month.

Kriging

Since the rainfall data appear to be spatially correlated, kriging may be a useful method for obtaining rainfall predictions at unsampled locations; kriging incorporates the variogram model of spatial correlation into the estimates. See section 4.3 for details on how to fit kriging models in S+SPA-TIALSTATS.

For this data, we use ordinary kriging to predict the correlated residuals over the same convex hull region that we used for the local regression predictions. We then add this to the large-scale trend modeled using local regression. The kriging results obtained using the S+SPATIALSTATS functions `krige` and `predict.krige` are shown in figure 7.8. If analysis of the data or knowledge of the operating processes justify an assumption of local stationarity, then ordinary kriging of the original sampled values using the variograms from the residuals, may also be appropriate.

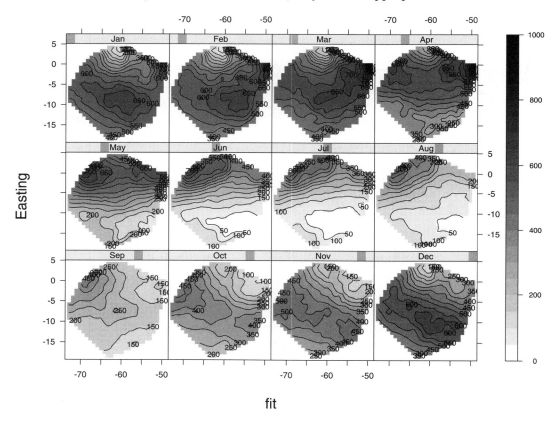

FIGURE 7.8. Kriging results for monthly rainfall. Displayed values are the sum of the predicted loess regression surface and the kriging predictions for the residuals.

7.2.5 Exporting Analysis Results to ARC/INFO

There are several results from the spatial analysis in S-PLUS that might be interesting to transfer back to ARC/INFO. Predictions using loess models yield fitted values and residuals. One (since the other can be calculated)

or both of these values can be exported to ARC/INFO as part of the
RAIN.PAT table, using the S+GISLINK command `splusarc`. Additionally,
the surfaces generated by kriging predictions, and their associated standard
errors, can be exported to ARC/INFO as grids, using the S+GISLINK
command `splusgrid`.

7.3   Example: Analysis of a Polygon Coverage

Polygon coverages in ARC/INFO will generally correspond to spatial data
that can be treated as irregular lattice data. Alternatively, methods generally associated with geostatistical data might be applied to data from a
polygon coverage. The example we analyze in this section will be treated
as irregular lattice data, as defined in section 1.1. The polygon data represent the regions associated with census tracts, for which a variety of
demographic data was collected.

Note: This data set is available as part of the S+GISLINK product.

7.3.1   LA Coverage

The ARC/INFO **LA** coverage is a polygon coverage of the census tracts
for the city of Los Angeles, California. It contains demographic information
such as income, burglary rates, and population densities for 667 census
tracts.

Maps of the **LA** coverage, generated in ARCPLOT using the command
POLYSHADES, suggest possible spatial structure in the occurrence of burglaries in Los Angeles. There seems to be a higher rate of burglaries in the
center of the city than to the north. Questions of interest one might want
to explore include: 1) can the burglary rates be explained by other demographic information, such as household income; and 2) do neighboring
tracts exhibit influence on one another. These types of questions can be
explored by transferring your data to S-PLUS and using analytical tools
available in S-PLUS and S+SPATIALSTATS.

7.3.2   Importing LA Data into S-PLUS

In preparation for importing the **LA** coverage into S-PLUS there are several things that should be considered. First, the coverage contains many
demographic variables, some of which may not be used in an analysis. Do
we want to transfer them all to S-PLUS? Second, the polygon label points

will be used to represent the locations of each region. Are the locations of the label points satisfactory for the proposed analyses? Third, spatial analysis of lattice data requires some sort of neighborhood information. How will we define neighbors for the census tract regions? The answers to these questions are addressed below. Use the following commands to prepare to import the **LA** data into S-PLUS.

```
Arc: copy la lanew
Arc: PULLITEMS LANEW.PAT LANEW.PAT

Enter item names (type END or a blank line when done):
===========================================================
Enter the 1st item: AREA
Enter the 2nd item: PERIMETER
Enter the 3rd item: LANEW#
Enter the 4th item: LANEW-ID
Enter the 5th item: CTRACT
Enter the 6th item: POPDENS
Enter the 7th item: CROWD
Enter the 8th item: YOUTH
Enter the 9th item: INCOME
Enter the 10th item: BURG.RATE
Enter the 11th item: END

Arc: addxy lanew
Arc: polyneigh lanew
```

As we did with the **rainfall** coverage (section 7.2.2), we have used the ARC COPY command to make a working copy of the **LA** coverage. The **LANEW** coverage contains 25 data items, many of which we will not use. The INFO PULLITEMS command is used to create a new LANEW.PAT table which contains the demographic items of interest, in addition to the four standard .PAT items AREA, PERIMETER, LANEW#, and LANEW-ID.

The ARC ADDXY command is used to add the x, y coordinates of each polygon label point to the new LANEW.PAT table. For this coverage, we do not know what the label points represent; they could be arbitrary points added while digitizing, or they could be the geometric centroids of the polygons. The use of geometric centroids is preferred; ideally, the user would recreate the label points, if necessary, before using ADDXY.

The ARC POLYNEIGH command is an AML, which is provided as part of S+GISLINK version 1.1, that creates a polygon neighbor list (based on RPOLY and LPOLY). The POLYNEIGH command above creates an INFO table of neighbor relationships (a Polygon Neighbor List): LANEW.PNL. Along with the neighbors for each polygon, LANEW.PNL contains the lengths of common boundaries between neighbors. The `.PNL` table is keyed to the `.PAT` table through the `LANEW#` item.

The **LANEW** coverage is now ready to be transferred to S-PLUS as a data frame using the S+GISLINK `arcsplus` command with the `poly` option. Enter the following commands to import the **LANEW** coverage and the `LANEW.PNL` INFO table into S-PLUS as data frames:

```
Arc: arcsplus lanew poly lapat
Arc: infosplus lanew.pnl la.pnl
```

The S+GISLINK `arcsplus` command copies the data from `LANEW.PAT` to the S-PLUS data frame `lapat`. The S+GISLINK `infosplus` command copies the data from `LANEW.PNL` to the S-PLUS data frame `la.pnl`. Use the S+GISLINK `splusexecute` command to view the items (columns) of the `lapat` and `la.pnl` data frames as follows:

```
Arc: splusexecute 'names(lapat)'
 [1] "AREA"       "PERIMETER"  "LA.NEW."    "LA.NEW.ID"
 [5] "CTRACT"     "POPDENS"    "CROWD"      "YOUTH"
 [9] "INCOME"     "BURG.RATE"  "X.COORD"    "Y.COORD"
Arc: splusexecute 'names(la.pnl)'
[1] "POLY."     "NEIGHBOR." "WEIGHT"
Arc: splusexecute 'length(la.pnl[,1])'
[1] 3666
Arc: splusexecute 'la.pnl[4:11,])'
   POLY. NEIGHBOR.     WEIGHT
 4    17        57   8.860026
 5    17        48  11.948385
 6    17        15  16.924734
 7    17        39   5.871436
 8    17        50   1.771038
 9    57        17  33.947208
10    57        48  23.410501
11    57        56  42.642288
```

The `lapat` data frame has 12 items. The `la.pnl` data frame has 3 items each with 3,666 records: `POLY.` occurs once for each neighbor relationship, `NEIGHBOR.` is a neighbor of the corresponding polygon in `POLY.`, and `WEIGHT` is the length of the common boundary between two polygons, which could be used to weigh the neighbor relationships. In the list shown above, polygon 17 has polygons 57, 48, 15, 39, and 50 as neighbors. Likewise, polygon 57 has neighbors 17, 48, and 56.

7.3.3   Removing Excess ARC/INFO Information

Since the **LA** data comes from a polygon coverage, we need to remove the universal polygon (which should be the first row) and any other extraneous data. Do this as follows:

1. Check for negative values of `AREA`, since this is a defining characteristic of the universal polygon.

   ```
   > la.index <- seq(along=lapat$AREA)
   > negarea <- lapat$AREA < 0
   > rows.negarea <- la.index[negarea]
   > rows.negarea
   [1] 1
   ```

2. Check for duplicate records in `lapat$LA.NEW.ID`.

   ```
   > lapat$LA.NEW.ID[duplicated(lapat$LA.NEW.ID)]
   [1] 0 0 0 0 0 0 0 0 0 0 0
   ```

   There are 12 rows that have a value of 0 in the `lapat$LA.NEW.ID` field. The remaining `lapat$LA.NEW.ID` values are unique.

3. Determine which rows have a 0 for `lapat$LA.NEW.ID`.

   ```
   > badid <- lapat$LA.NEW.ID == 0
   > rows.badid <- la.index[badid]
   > rows.badid
    [1]    1   12 220 237 271 364 479 634 636 637 638 659
   ```

   Notice that row 1, the universal polygon, is included in `rows.badid`.

4. Check the data in the remaining rows for records with bad identifiers, to determine if we would lose valuable information should these rows be removed.

   ```
   > lapat[rows.badid[-1],5:11]
       CTRACT POPDENS CROWD YOUTH INCOME BURG.RATE
    12      0       0    -1     0      0        -1
   220      0       0    -1     0      0        -1
   237      0       0    -1     0      0        -1
   271      0       0    -1     0      0        -1
   364      0       0    -1     0      0        -1
   479      0       0    -1     0      0        -1
   634      0       0    -1     0      0        -1
   636      0       0    -1     0      0        -1
   ```

637	0	0	-1	0	0	-1
638	0	0	-1	0	0	-1
659	0	0	-1	0	0	-1

The data values are all 0's or -1's indicating extraneous data records.

5. Remove the 12 bad rows of data while making a working copy of the data frame.

```
> la <- lapat[!badid,]
> dim(la)
[1] 655  13
```

The new S-PLUS object `la` has 655 records and 13 variables. We are now ready to proceed with a spatial analysis of these data.

7.3.4 Exploratory Data Analysis

The predictors that might be useful in modeling of burglary rates include population density, crowding, number of youth, household income, and high-school dropout rate. Use the S-PLUS `hist` function to look at the distributions of each variable:

```
> par(mfrow=c(3,2))
> for (i in 7:11) {
+       hist(la[,i],nclass=25,xlab=names(la)[i])
+ }
```

The resulting histograms are shown in figure 7.9. Due to the skewness of most of the variables, a log-transform was applied to the data values to obtain more normally distributed data. The resulting histograms are shown in figure 7.10. Based on these histograms, the log-transformation seems a reasonable choice for every variable except YOUTH.

The relationship between burglary rates and possible predictors can be visualized with scatterplots created using the S-PLUS `plot` function. Scatterplots for burglary rates as a function of income, population density, percentage of youth, and an index of crowding are shown in figure 7.11. The lines on the plots represent the fit from a simple linear model, found and plotted using the S-PLUS `lm` and `abline` functions.

The scatterplots show a negative relationship between burglary rates and household income, and positive relationships for the other variables. The strongest relationship seems to be with income.

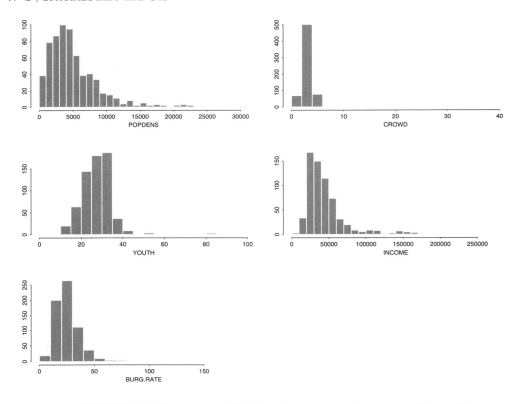

FIGURE 7.9. Histograms of LA burglary rates and some possible predictors.

To look at the spatial distribution of the burglary rate, we can break the data into groups based on the original histogram in figure 7.9, and plot the groups relative to the locations of the census tracts:

```
> br.groups <- cut(la$BURG.RATE, breaks=c(10,30,50))
> br.groups <- as.numeric(br.groups)
> scaled.plot(la$X.COORD, la$Y.COORD, type="n")
> text(la$X.COORD, la$Y.COORD, br.groups)
```

The resulting plot is shown in figure 7.12. From this plot there appears to be higher burglary rate between the y coordinates 3752286 and 3775729.

Another way to look for spatial structure in the data is to test for spatial autocorrelation. In S+SPATIALSTATS the function `spatial.cor` can be used to calculate Moran or Geary spatial correlation coefficients. Calculate the Moran coefficient for the log-transformed burglary rates as follows:

1. Create an object of class `"spatial.neighbor"` (section 5.1.1) using the data in `la.pnl`. Then use the S+SPATIALSTATS `spatial.subset`

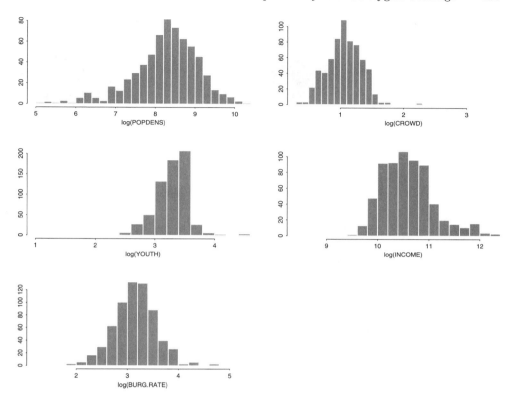

FIGURE 7.10. Histograms of log-transformed LA burglary rates and some possible predictors.

function to remove the neighbor relationships for the 12 bad records identified in section 7.3.2.

```
> la.snhbr <- spatial.neighbor(row.id = la.pnl$POLY.,
+       col.id = la.pnl$NEIGHBOR,
+       weights = la.pnl$WEIGHT)
> la.tnhbr <- spatial.subset(neighbor = la.snhbr,
+       region.id = 1:667, subset = !badid,
+       reorder = T)
> la.snhbr <- la.tnhbr$neighbor
> la.snhbr[1:8,]
Total number of spatial units =  655
(Matrix was NOT defined as symmetric)
    row.id col.id   weights matrix
  4     15     55  8.860026      1
  5     15     46 11.948385      1
```

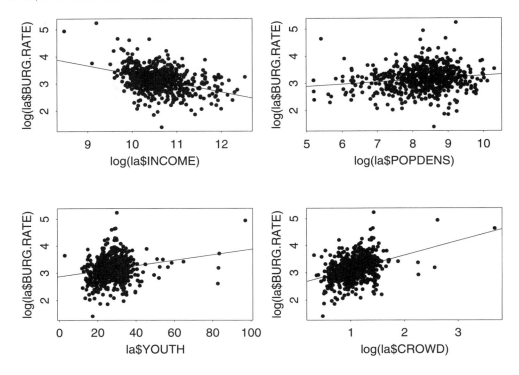

FIGURE 7.11. Scatterplots of burglary rates as a functions of possible predictors.

6	15	13	16.924734	1
7	15	37	5.871436	1
8	15	48	1.771038	1
9	55	15	33.947208	1
10	55	46	23.410501	1
11	55	54	42.642288	1

The columns of the spatial neighbor object contain essentially the same information as the data in `la.pnl`.

2. Call `spatial.cor` to calculate the Moran coefficient.

```
> br.moran <- spatial.cor(log(la$BURG.RATE),
+       la.snhbr, statistic="moran")
> br.moran

        Spatial Correlation Estimate

Statistic = "moran" Sampling = "nonfree"
```

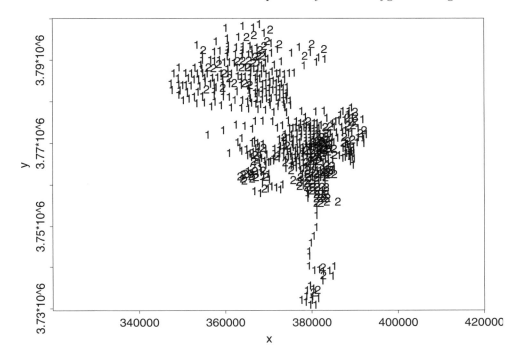

FIGURE 7.12. Census tract locations showing 3 groupings for burglary rates.

```
Correlation =  0.4522
Variance    =  0.001656
Std. Error  =  0.04069

Normal statistic =  11.15
Normal p-value (2-sided) =  7.091e-29

Null Hypothesis:  No spatial autocorrelation
```

The resulting Moran coefficient is 0.4522, indicating the presence of positive spatial autocorrelation, given our choice of neighbors and neighbor weights; the p-value from an approximate test using Gaussian theory is essentially 0. The spatial autocorrelation may be due to trend or to an underlying stochastic process. See section 5.2 for additional details on spatial autocorrelation.

We can also look for spatial autocorrelation in the residuals from a simple linear model that uses log-transformed household income as a predictor. This can be done as follows:

1. Model log-transformed burglary rates as a function of log-transformed household income and calculate the residuals from this fit.

```
> br.lm <- lm(log(BURG.RATE) ~ log(INCOME), data=la)
> br.lmres <- residuals(br.lm)
```

2. Calculate the Moran coefficient on the linear model residuals.

```
> br.lmres.moran <- spatial.cor(br.lmres,
+        la.snhbr,statistic="moran")
> br.lmres.moran
```

```
        Spatial Correlation Estimate

Statistic = "moran" Sampling = "nonfree"

Correlation =  0.2765
Variance    =  0.001657
Std. Error  =  0.04071

Normal statistic =  6.83
Normal p-value (2-sided) =  8.494e-12

Null Hypothesis:  No spatial autocorrelation
```

The Moran statistic still suggests the presence of spatial autocorrelation versus the null hypothesis of no spatial autocorrelation. There might also be some leftover trend that was not captured by the model.

Note: The use of the Moran statistic on residuals from a linear model is not necessarily valid (Ripley, 1981, p. 99). The results are presented here solely for illustrative purposes.

7.3.5 Spatial Linear Model

One approach to modeling the burglary rate is to use a spatial linear model. This is a linear model that accounts for the spatial structure in the data. Using a spatial linear model, we can attempt to fit the burglary rates by including explanatory variables, along with small-scale spatial variation due to interactions with neighbors. In this case we will use the previously created spatial neighbor object la.snhbr, which defines neighbors as all adjacent census tracts. The neighbors are weighted by the length of their shared boundaries. The S+SPATIALSTATS slm function can be used to model log-transformed burglary rates, as follows:

```
> br.slm <- slm(log(BURG.RATE) ~ log(INCOME),
+      data=la, cov.family=SAR,
+      spatial.arglist=list(neighbor=la.snhbr))
```

For this model we have specified a simultaneous spatial autoregression
(SAR) covariance family. This defines the form of the model covariance.
The SAR model was chosen over a conditional spatial autoregressive model
(CAR) because the use of CAR requires a symmetric dispersion matrix. See
section 5.3.1 for a disscusion on covariance families.

The results of the model fit are displayed using the **summary** function:

```
> summary(br.slm)
Call:
slm(formula = log(BURG.RATE) ~ log(INCOME),
    cov.family = SAR, data = la,
    spatial.arglist = list(neighbor = la.snhbr))
Residuals:
    Min     1Q   Median    3Q    Max
 -1.713 -0.2248 0.001997 0.2164 1.578


Coefficients:
              Value Std. Error  t value Pr(>|t|)
(Intercept)  6.3669    0.3817   16.6809   0.0000
log(INCOME) -0.3046    0.0359   -8.4868   0.0000


Residual standard error: 0.377476 on
          652 degrees of freedom


Variance-Covariance Matrix of Coefficients
            (Intercept)  log(INCOME)
(Intercept)  0.14568732 -0.013684063
log(INCOME) -0.01368406  0.001288399


Correlation of Coefficient Estimates
            (Intercept) log(INCOME)
(Intercept)  1.0000000   -0.9988011
log(INCOME) -0.9988011    1.0000000


rho =   8.869e-5


Iterations = 5
Gradient norm =  0.01024
Log-likelihood =  -1497


Convergence:  RELATIVE FUNCTION CONVERGENCE
```

The generic **summary** function used above provides information on the co-efficient estimates. The coefficient estimates for the linear component are 6.3669 for the grand mean and -0.3046 for log-transformed household income. The coefficient estimates are similar to those returned from the linear model (grand mean = 6.552086, log(INCOME) = -0.3212198) that was originally fitted to the data. The given coefficient estimates imply that as household income goes up, burglary rates go down. This is consistent with the top left scatterplot in figure 7.11.

The standard errors for the coefficients are given, along with the associated t-statistics. Both coefficients are significantly different from zero, based on the two-tailed p-values given. The summary statistics shown for residuals are for scaled residuals if a scaling was used (see section 5.3.3). For this example, the residuals are unscaled. The estimate for ρ is 8.869e-5; ρ is the parameter by which the neighbor matrix is multiplied to obtain the covariance matrix. See section 5.3.2 for a complete discussion of the use of **slm**.

7.3.6 Model Selection

The spatial linear model developed in the previous section is only one possible model for the burglary rates. In addition to the choice of explanatory variables, the choice of neighbors, neighbor weights, and covariance family may also significantly affect model outcome. Section 5.3.3 reviews some model comparison and residual diagnostic techniques available in S-PLUS and S+SPATIALSTATS. In this section we look at a couple of model selection criteria.

First, we can use a likelihood ratio test to determine if the parameter estimate for ρ is significantly different from zero:

```
> lrt(br.slm, parameters=0)
Likelihood Ratio Test

Chisquare statistic = 42.3182, df = 1,
      p.value = 7.756751e-11

parameters:
 param1(fixed)
            0

coefficients:
 (Intercept) log(INCOME)
   6.552086   -0.3212198
```

Although the ρ estimate from the model appears to be very small (8.869e-5), the estimate is significantly different from zero; the p-value is essentially 0. The ρ estimate is probably so small due to the large value of the neighbor weights, where the weights are the lengths of shared boundaries. A more reasonable ρ estimate might be obtained by scaling the weights for each set of neighbors by the total boundary lengths.

Using slm, the variation in burglary rates has been modeled with a linear household income component, a neighbor covariance component (termed the "signal" (Haining, 1990, p. 258)), and a residual variation component. In a spatial linear model, the predicted values are computed as the fitted values from the linear component plus the signal. Calculate the signal component and plot the predicted burglary rates versus the actual rates (see section 5.3.4 for details):

```
> br.signal <- (log(la$BURG.RATE) - fitted(br.slm)
+        - resid(br.slm))
> xmax <- max(log(la$BURG.RATE))
> ymax <- max(fitted(br.slm) - resid(br.slm))
> plot(log(la$BURG.RATE), fitted(br.slm) + br.signal,
+      xlim=c(1,xmax),ylim=c(1,ymax))
> abline(0,1)
```

The resulting plot is shown in figure 7.13. The line on the plot represents the points at which the predicted values equal the observed values. The variation in the burglary rates is not completely captured by the current model. More explanatory variables are probably needed.

7.3.7   Exporting Analysis Results to ARC/INFO

There are several results from spatial modeling in S-PLUS that might be interesting to transfer to ARC/INFO. The spatial linear model resulted in fitted values for the linear model component and predicted values that incorporated the signal, or spatial neighborhood component. The fitted and predicted values could be exported back to ARC/INFO as part of the .PAT table, using the S+GISLINK splusarc command. A visual interpretation of the model would then be possible using colored maps showing predictions versus the data. See the S+SPATIALSTATS example in section 5.3.4 for consideration.

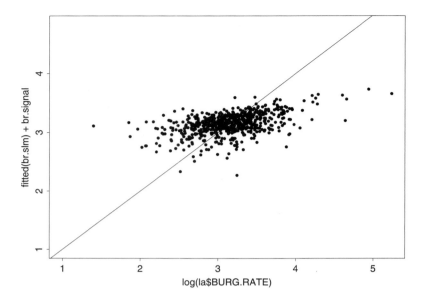

FIGURE 7.13. Predicted model values versus observed values for log-transformed burglary rates.

7.4   Example: Analysis of a Grid

Grids in ARC/INFO will generally correspond to spatial data that can be treated as gridded geostatistical data or as regular lattice data. The example we analyze in this section will be treated as regular lattice data. The grid represents regions where budworm infestation has been monitored.

7.4.1   Budworm Grid

The ARC/INFO **budworm** grid contains a classification value for budworm infestation for points on a 70×325 grid. Budworm infestation is classified on a scale of 0 to 8, where the extremes are 0 for no infestation to 8 for a large degree of infestation.

Note: This data set is not provided as part of S+SPATIALSTATS.

Exploratory mapping of the **budworm** grid, using the S-PLUS `image` function, shows definite clusters of infestation. Given the existing clusters, one question of interest might be: is there spatial autocorrelation within the individual clusters? An answer to this question can be explored using analytical tools available in S-PLUS and S+SPATIALSTATS.

Figure 7.14 shows a grayscale plot of the budworm grid. The lines on the plot show the boundaries for the larger cluster in the middle, which we will use to illustrate some of the techniques in S+SPATIALSTATS.

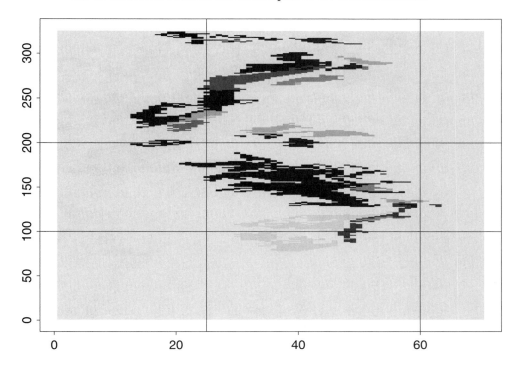

FIGURE 7.14. Grayscale plot for budworm infestation grid.

Extracting and plotting the larger middle cluster, as shown below, yields the plot shown in figure 7.15:

```
> section <- budworm[25:60, 100:200]
> image(section)
```

The new section is a 36 × 101 grid. The `neighbor.grid` function can be used to generate a set of first order neighbors for the grid:

```
> ngrid <- neighbor.grid(nrow=36, ncol=101)
> class(ngrid)
[1] "spatial.neighbor" "data.frame"
> ngrid[1:10,]
Total number of spatial units =  3636
(Matrix was NOT defined as symmetric)
```

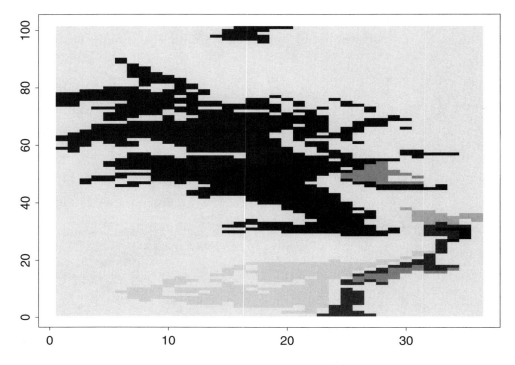

FIGURE 7.15. Grayscale plot for a large cluster of budworm infestation (subsection of figure 7.14).

	row.id	col.id	weights	matrix
1	1	2	1	1
2	1	37	1	1
3	2	1	1	1
4	2	3	1	1
5	2	38	1	1
6	3	2	1	1
7	3	4	1	1
8	3	39	1	1
9	4	3	1	1
10	4	5	1	1

The object ngrid is an object of class "spatial.neighbor". Using the default parameters for the call to neighbor.grid results in equally weighted neighbors defined by adjacent points on the grid (first order).

The S+SpatialStats function spatial.cor can be used to calculate a coefficient of spatial autocorrelation for the subset of the budworm grid. Use the spatial neighbor object created above to calculate a Geary autocorrelation coefficient:

```
> bw.geary <- spatial.cor(as.vector(section),
+       neighbor = ngrid, statistic = "geary",
+       npermutes = 100)
> bw.geary

        Spatial Correlation Estimate

Statistic = "geary" Sampling = "nonfree"

Correlation =   0.2765
Variance    =   0.008824
Std. Error  =   0.09393

Normal statistic =  -7.703
Normal p-value (2-sided) =  1.334e-14

Null Hypothesis:  No spatial autocorrelation

Quantiles of the permutation-correlations :
   Min. 1st Qu. Median   Mean 3rd Qu.  Max.
 0.9682  0.9921      1 0.9998   1.008 1.033

permutation p-value = 0
```

In this call to `spatial.cor`, the optional `npermutes` argument is specified as 100. This causes Geary coefficients to be calculated for 100 random permutations of the budworm data vector. The test for significance is then based on the resulting permutation distribution. The Geary coefficient of 0.2765 is significant when compared to the null hypothesis of no spatial autocorrelation; the permutation p-value is 0. The small (< 1) Geary coefficient indicates positive spatial autocorrelation within the cluster of budworm infestation. See section 5.2 for more information on the Geary coefficient and spatial autocorrelation.

The results of the test for spatial autocorrelation are consistent with the spatial structure apparent in the grayscale plot of figure 7.15.

Appendix A

Installing S+SPATIALSTATS

Installing S+SPATIALSTATS is a simple process, but the details differ depending on whether you are installing on a UNIX or Windows system. This chapter contains detailed procedures for both systems. If you have any difficulty with the installation which you cannot resolve from this chapter, please contact the S-PLUS Technical Support department at **(206) 283–8802**, ext. **235**, or send e-mail to **support@statsci.com**.

A.1   Installing S+SPATIALSTATS in UNIX

Installation of S+SPATIALSTATS on UNIX systems has two steps: reading the tape and installing the license for S+SPATIALSTATS. The installation requires write permission in the directory **Splus SHOME**.

A.1.1   Requirements

The S+SPATIALSTATS module requires S-PLUS version 3.3 or later. Check the version of S-PLUS you are using by typing the following at your shell prompt:

Splus VERSION

The output will look something like the following:

S-PLUS Version 3.3 Release 1 for Sun SPARC, SunOS 4.1.x:1995
Tape ID S4015Q written Thu Dec 16 15:00:00 PST 1995
Based on AT&T S VERSION Thu Apr 30 09:54:11 EDT 1992

If you do not have the correct version, you cannot install S+SPATIALSTATS. Call or fax MathSoft to obtain the new version.

A.1.2   Installation Procedure

To install S+SPATIALSTATS initially, use the following procedure. If you need to update your S+SPATIALSTATS license, either because you have reinstalled S+SPATIALSTATS or you are adding new users or a new host, see section A.1.4.

1. On a machine where S-PLUS is installed, type the following command at the shell prompt:

   Splus LICENSE server host

   The machine name returned is the license server host, the machine where the license server runs. You must be on this machine to get the correct host info number (step 3) and install the S+SPATIALSTATS license (step 5).

2. Log on to the license server host (see step 1). Verify the location of the S-PLUS installation by the command:

   Splus SHOME

3. Obtain the proper license key for S+SPATIALSTATS by faxing the MathSoft keycode department. Request a keycode for feature code S+SPATIALSTATS. You will need to obtain the host info number from the command **HOSTINFO** to complete the request form. To get this number, use the following commands:

   Splus LICENSE hostinfo

   The output will look something like:

   The server code that follows needs to be provided to your StatSci representative in order to produce a license key for use of this product. This should be run on the system that will be running the license management server daemon— the system on which the command
   **   Splus LICENSE server start**
   will be run.
   Code for server thyme is: 2098 2796 9021 6735 5

4. Set your current directory to the directory listed from the command **Splus SHOME**. Put the S+SpatialStats tape in a tape drive, and read the tape with one of the following commands.

Note: You must issue the following **tar** command from the **Splus SHOME** directory.

- If the license server host machine is the machine where you have loaded the tape, use the command:

  tar xvf /dev/rst0

  where **rst0** is the name of the local drive. Alternate names for the drives are often **/dev/rst1** or **/dev/rst8**.

- If the license server host machine is not the same as the machine where you have loaded the tape, use the following command:

  rsh *remotename* **-n dd if=/dev/rst0 | tar xBf -**

  where *remotename* is the name of the machine with the tapedrive and **rst0** is the local name of the tape drive on the remote machine. Alternate names for the drives are often **/dev/rst1** or **/dev/rst8**.

5. Install the keycode for S+SpatialStats obtained from MathSoft, using the following command:

   Splus MODINSTALL spatial

   You will be asked to enter the keycode you obtained from MathSoft.

   Note: You must be running on the machine which runs the license server to install the keycode.

A.1.3   Verifying Installation

Verify that the license key codes have been installed correctly, by typing:

Splus LICENSE features

This will generate a table of the features with installed license keys. For an installation with S-Plus and S+SpatialStats, this table should contain at least the following entries:

Feature Name	Alias
00	S-PLUS
12	S+SPATIALSTATS

To verify the operation of S+SPATIALSTATS do the following:

1. Start an S-PLUS session with the command:

   Splus

2. In S-PLUS, look at the help file for S+SPATIALSTATS by typing:

   ```
   > module(help="spatial")
   ```

3. Load the S+SPATIALSTATS module with the command:

   ```
   > module(spatial)
   ```

4. Try a simple S+SPATIALSTATS command:

   ```
   > lansing.spp <- spp(lansing)
   ```

A.1.4   Updating Your S+SPATIALSTATS License

If you reinstall S+SPATIALSTATS, or need to modify your license to support more users or run on a different host, use the **Splus LICENSE** utility as follows:

Splus LICENSE modinstall spatial

You are prompted for the license keycode.

Note: If you change hosts or the number of users, you must contact MathSoft for a new license keycode.

A.2   Installing S+SPATIALSTATS in Windows

Installing S+SPATIALSTATS on a Windows system is essentially a one step process: you run the Setup program included on the S+SPATIALSTATS distribution disk. Help is available for each of the main dialog boxes in the setup procedure. Select the Help button for information about the current dialog box. To leave Setup before the setup is complete, select Exit from any Setup dialog box: Setup displays the S+SPATIALSTATS Setup Exit dialog to remind you that you have not completed the setup procedure.

To set up S+SPATIALSTATS on a Windows system:

1. Start Microsoft Windows if it is not already running.

2. Insert the S+SPATIALSTATS disk in the appropriate drive, A or B.

3. Choose Run from the File menu of the Program Manager (Windows 3.1), or Run from the Start menu (Windows 95).

4. In the Command Line text field of the Run dialog, type **A:\SETUP**, and then press $\boxed{\text{Enter}}$ or click the OK button. (If your disk drive is drive **B:**, replace **A:** with **B:** in the above command.)

   Setup displays a message that it is initializing itself. When initialization is complete, a dialog box welcomes you to the Setup program.

5. Click the button labeled Continue. Setup searches for a directory containing S-PLUS for Windows version 3.3 by checking your DOS environment and the file **SPLUS.INI** for the environment variable **SHOME**. If Setup finds a valid **SHOME**, it displays a dialog indicating the **SHOME** for the default installation directory. The dialog also lists the two installation components—the Runtime Files and the Help Files—together with the approximate disk space required to install them in the installation directory.

   If Setup does *not* find a valid **SHOME**, it displays a dialog box, with the default directory name **C:\SPLUSWIN** appearing highlighted. Type the complete path of a directory containing S-PLUS for Windows version 3.3, then press $\boxed{\text{Enter}}$ or click OK.

6. (*Optional Step.*) You can specify a different installation directory, if the alternate directory also contains S-PLUS for Windows version 3.3.

   To specify a different directory for the installation, click the button labeled Change Installation Dir... from the dialog labeled S+SPATIALSTATS for Windows Setup Options. A dialog box appears, labeled Change S-PLUS SpatialStats Installation Directory. Type the path specification for any directory containing S-PLUS for Windows version 3.3 in the text field labeled Directory, then click OK.

   The specified directory should be on a drive with at least as much available disk space as specified in the Space Required field. (Approximately 2.5Mb for the Runtime Files, 0.5MB for the Help Files.)

7. Select which files to install. By default, both the Runtime Files and Help Files are installed. Only the Runtime Files are required to use S+SPATIALSTATS.

The options selected for installation are indicated with an X in the associated box. The disk space required to install each selected option is also indicated. Click on a box to select or unselect an option as needed.

8. When you are satisfied with the installation directory and the installation options, click Continue. Setup installs the selected files to the appropriate subdirectories of the specified installation directory. When all files are installed, Setup displays a dialog box requesting a serial number.

9. Enter your serial number (found on your S+SPATIALSTATS floppy disk) in the text field labeled Serial No. The serial number is the seven-character string beginning with "ST."

When the setup is complete, Setup displays a dialog box indicating Setup is complete.

Now read chapter 2 to get started using S+SPATIALSTATS. If you are a novice S-PLUS user with limited experience working in DOS or Windows, turn to the booklet, *A Gentle Introduction to* S-PLUS. If you are an experienced computer user with a strong background in data analysis, turn to the booklet, *A Crash Course in* S-PLUS.

Appendix B

S+SPATIALSTATS Functions and Data Sets by Category

This appendix gives a list of functions and utilities for S+SPATIALSTATS by category. Documentation for these datasets and functions is available both in the help files in this manual and on-line using the `help` function.

Add to Existing Plot

`hex.legend`	Add a Legend Hexagonal Lattice Plot
`hexagons`	Add Hexagonal Cells to Plot of `"hexbin"` Object
`identify.hexbin`	Identify Points On a Hexagonal Binned Plot
`identify.vgram.cloud`	Identify Points On a Variogram Cloud Plot

Computations Related to Plotting

`boxplot.vgram.cloud`	Boxplots of a Variogram Cloud Object

Data Sets

`aquifer`	Wolfcamp Aquifer Data
`bramble`	Bramble Cane Data
`coal.ash`	Coal Ash Data
`iron.ore`	Iron Ore Data
`lansing`	Lansing Woods Tree Data
`quakes.bay`	Bay Area Earthquakes
`quakes.wash`	Washington State Earthquakes In 1980
`scallops`	Scallop Abundance Data
`sids`	Sudden Infant Death Syndrome Data 1974-1978
`sids.neighbor`	Neighbors for Sudden Infant Death Syndrome Data
`sids2`	Sudden Infant Death Syndrome Data 1979-1984
`wheat`	Wheat Grain and Straw Yield

Geostatistical Data Analysis

anisotropy.plot	Explore Corrections For Geometric Anisotropy
boxplot.vgram.cloud	Boxplots of a Variogram Cloud Object
correlogram	Empirical Correlogram and Covariogram
covariogram	Empirical Correlogram and Covariogram
exp.cov	Theoretical Distance Based Covariance Functions
exp.vgram	Theoretical Variogram Functions
gauss.cov	Theoretical Distance Based Covariance Functions
gauss.vgram	Theoretical Variogram Functions
identify.vgram.cloud	Identify Points On a Variogram Cloud Plot
krige	Ordinary and Universal Kriging
linear.vgram	Theoretical Variogram Functions
loc	Correct Spatial Locations for Geometric Anisotropy
model.variogram	Display a Variogram Object and Theoretical Model
plot.correlogram	Plot a Variogram, Covariogram or Correlogram
plot.covariogram	Plot a Variogram, Covariogram or Correlogram
plot.variogram	Plot a Variogram, Covariogram or Correlogram
power.vgram	Theoretical Variogram Functions
predict.krige	Ordinary and Universal Kriging Prediction
spher.cov	Theoretical Distance Based Covariance Functions
spher.vgram	Theoretical Variogram Functions
woway.formula	Fit of a Two-Way Table (Formula Method)
variogram	Empirical Variogram
variogram.cloud	Calculate Variogram Cloud

Hexagonal Binning

cell2xy	Compute x,y Coordinates From Hexagon Cell Ids
erode.hexbin	Erode a Hexagonally Binned Image
hex.legend	Add a Legend Hexagonal Lattice Plot
hexagons	Add Hexagonal Cells to Plot of "hexbin" Object
hexbin	Bivariate Binning into Hexagonal Cells
identify.hexbin	Identify Points On a Hexagonal Binned Plot
plot.hexbin	Plot A Hexagonal Lattice
rayplot	Adds Rays with Optional Confidence Arcs (Sectors)
smooth.hexbin	Hexagonal Bin Smoothing
summary.hexbin	Summary Method for a Hexagonally Binned Object
xy2cell	Compute Hexagon Cell Ids From x and y

High-Level Plots

anisotropy.plot	Explore Corrections For Geometric Anisotropy
boxplot.vgram.cloud	Boxplots of a Variogram Cloud Object
plot.correlogram	Plot a Variogram, Covariogram or Correlogram
plot.covariogram	Plot a Variogram, Covariogram or Correlogram
plot.hexbin	Plot A Hexagonal Lattice
plot.spp	Plots an Object of class spp
plot.variogram	Plot a Variogram, Covariogram or Correlogram
plot.vgram.cloud	Plot a Variogram Cloud
rayplot	Adds Rays with Optional Confidence Arcs (Sectors)
scaled.plot	Equal Scales Plot

Input/Output–Files
read.geoeas	Read A GEO-EAS Data File
read.neighbor	Read ASCII Files Containing Spatial Contiguity Information
write.geoeas	Write A GEO-EAS Data File

Interacting with Plots
identify.hexbin	Identify Points On a Hexagonal Binned Plot
identify.vgram.cloud	Identify Points On a Variogram Cloud Plot

Multivariate Techniques
find.neighbor	Find the Nearest Neighbors of a Point
quad.tree	Order a Multicolumn Real Matrix into a Quad Tree
twoway	Fit of a Two-Way Table
twoway.default	Fit of a Two-Way Table

Point Pattern Analysis
as.spp	Spatial Point Pattern Objects
bbox	Bounding Box for a Spatial Point Pattern
Fhat	EDF of Origin-to-Point Nearest Neighbor Distances
Ghat	EDF of Point-to-Point Nearest Neighbor Distances
intensity	Estimate the Intensity of a Spatial Point Pattern
is.spp	Spatial Point Pattern Objects
Kenv	Compute Simulations of Khat
kern2d	Kernel Smoothing of a Two-Dimensional Density
Khat	Ripley's K Function for a Spatial Point Pattern Object
Lenv	Compute Simulations of Khat
Lhat	Ripley's K Function for a Spatial Point Pattern Object
make.pattern	Generate a Spatial Point Process
plot.spp	Plots an Object of class spp
points.in.poly	Find Points Inside a Given Polygon
poly.area	Computes the Area of a Polygon
poly.grid	Generate a Grid Inside a Given Polygonal Boundary
scaled.plot	Equal Scales Plot
spp	Spatial Point Pattern Objects
summary.spp	Summary Method for a Spatial Point Pattern Object
triangulate	Delaunay's Triangulation

Printing
print.slm	Use print() on a slm Object
print.spatial.cor	Use print() on a spatial.cor Object
print.spatial.neighbor	Print a spatial.neighbor Object
print.summary.slm	Use print() on a summary.slm Object

Regression
slm	Fit a Spatial Linear Regression Model
summary.slm	Summary Method for Spatial Linear Models

Robust/Resistant Techniques
twoway	Fit of a Two-Way Table
twoway.default	Fit of a Two-Way Table
twoway.formula	Fit of a Two-Way Table (Formula Method)

Spatial Regression

CAR	Conditional Spatial Autoregression Object
check.islands	Detect Isolated Spatial Regions
cov.family.object	Spatial Family Object
find.neighbor	Find the Nearest Neighbors of a Point
lrt.slm	Likelihood Ratio Test for Spatial Linear Models
MA	Moving Average Spatial Regression Object
neighbor.grid	Creates a "spatial.neighbor" Object for a Grid
print.slm	Use print() on a slm Object
print.spatial.cor	Use print() on a spatial.cor Object
print.spatial.neighbor	Print a spatial.neighbor Object
print.summary.slm	Use print() on a summary.slm Object
quad.tree	Order a Multicolumn Real Matrix into a Quad Tree
read.neighbor	Read ASCII Files Containing Spatial Contiguity Information
SAR	Simultaneous Spatial Autoregression Object
slm	Fit a Spatial Linear Regression Model
slm.fit	Fitting Function for Spatial Linear Models
slm.fit.spatial	Fit a Spatial Linear Model (Generalized Least Squares)
slm.nlminb	Fit a Profile Likelihood in a Spatial Regression Model
spatial.cg.solve	Solve S b = x
spatial.condense	Remove Redundancy in "spatial.neighbor" Objects
spatial.cor	Measures of Spatial Correlation
spatial.determinant	Compute Sparse Matrix Determinant
spatial.multiply	Compute Sparse Matrix Vector Product A x
spatial.neighbor	Create a "spatial.neighbor" Object
spatial.neighbor.object	Class "spatial.neighbor"
spatial.solve	Solve S b = x
spatial.subset	Subset an Object of Class "spatial.neighbor"
spatial.sum	Sum of Two Objects of Class "spatial.neighbor"
spatial.weights	Compute a Spatial Weight Matrix
summary.slm	Summary Method for Spatial Linear Models

Spatial Statistics Module

anisotropy.plot	Explore Corrections For Geometric Anisotropy
aquifer	Wolfcamp Aquifer Data
as.spp	Spatial Point Pattern Objects
bbox	Bounding Box for a Spatial Point Pattern
boxplot.vgram.cloud	Boxplots of a Variogram Cloud Object
bramble	Bramble Cane Data
CAR	Conditional Spatial Autoregression Object
cell2xy	Compute x,y Coordinates From Hexagon Cell Ids
check.islands	Detect Isolated Spatial Regions
coal.ash	Coal Ash Data
correlogram	Empirical Correlogram and Covariogram

Spatial Statistics Module, (cont.)

cov.family.object	Spatial Family Object
covariogram	Empirical Correlogram and Covariogram
erode.hexbin	Erode a Hexagonally Binned Image
exp.cov	Theoretical Distance Based Covariance Functions
exp.vgram	Theoretical Variogram Functions
Fhat	EDF of Origin-to-Point Nearest Neighbor Distances
find.neighbor	Find the Nearest Neighbors of a Point
gauss.cov	Theoretical Distance Based Covariance Functions
gauss.vgram	Theoretical Variogram Functions
Ghat	EDF of Point-to-Point Nearest Neighbor Distances
hex.legend	Add a Legend Hexagonal Lattice Plot
hexagons	Add Hexagonal Cells to Plot of "hexbin" Object
hexbin	Bivariate Binning into Hexagonal Cells
identify.hexbin	Identify Points On a Hexagonal Binned Plot
identify.vgram.cloud	Identify Points On a Variogram Cloud Plot
intensity	Estimate the Intensity of a Spatial Point Pattern
iron.ore	Iron Ore Data
is.spp	Spatial Point Pattern Objects
Kenv	Compute Simulations of Khat
kern2d	Kernel Smoothing of a Two-Dimensional Density
Khat	Ripley's K Function for a Spatial Point Pattern Object
krige	Ordinary and Universal Kriging
lansing	Lansing Woods Tree Data
Lenv	Compute Simulations of Khat
Lhat	Ripley's K Function for a Spatial Point Pattern Object
linear.vgram	Theoretical Variogram Functions
loc	Correct Spatial Locations for Geometric Anisotropy
lrt	Computes a Likelihood Ratio Test
lrt.slm	Likelihood Ratio Test for Spatial Linear Models
MA	Moving Average Spatial Regression Object
make.pattern	Generate a Spatial Point Process
model.variogram	Display a Variogram Object and Theoretical Model
neighbor.grid	Creates a "spatial.neighbor" Object for a Grid
plot.correlogram	Plot a Variogram, Covariogram or Correlogram
plot.covariogram	Plot a Variogram, Covariogram or Correlogram
plot.hexbin	Plot A Hexagonal Lattice
plot.spp	Plots an Object of class spp
plot.variogram	Plot a Variogram, Covariogram or Correlogram
plot.vgram.cloud	Plot a Variogram Cloud
points.in.poly	Find Points Inside a Given Polygon
poly.area	Computes the Area of a Polygon
poly.grid	Generate a Grid Inside a Given Polygonal Boundary
power.vgram	Theoretical Variogram Functions
predict.krige	Ordinary and Universal Kriging Prediction
print.slm	Use print() on a slm Object

Spatial Statistics Module, (cont.)

print.spatial.cor	Use print() on a spatial.cor Object
print.spatial.neighbor	Print a spatial.neighbor Object
print.summary.slm	Use print() on a summary.slm Object
quad.tree	Order a Multicolumn Real Matrix into a Quad Tree
quakes.bay	Bay Area Earthquakes
quakes.wash	Washington State Earthquakes In 1980
rayplot	Adds Rays with Optional Confidence Arcs (Sectors)
read.geoeas	Read A GEO-EAS Data File
read.neighbor	Read ASCII Files Containing Spatial Contiguity Information
SAR	Simultaneous Spatial Autoregression Object
scaled.plot	Equal Scales Plot
scallops	Scallop Abundance Data
sids	Sudden Infant Death Syndrome Data 1974-1978
sids.neighbor	Neighbors for Sudden Infant Death Syndrome Data
sids2	Sudden Infant Death Syndrome Data 1979-1984
slm	Fit a Spatial Linear Regression Model
slm.fit	Fitting Function for Spatial Linear Models
slm.fit.spatial	Fit a Spatial Linear Model (Generalized Least Squares)
slm.nlminb	Fit a Profile Likelihood in a Spatial Regression Model
smooth.hexbin	Hexagonal Bin Smoothing
spatial.cg.solve	Solve S b = x
spatial.condense	Remove Redundancy in "spatial.neighbor" Objects
spatial.cor	Measures of Spatial Correlation
spatial.determinant	Compute Sparse Matrix Determinant
spatial.multiply	Compute Sparse Matrix Vector Product A x
spatial.neighbor	Create a "spatial.neighbor" Object
spatial.neighbor.object	Class "spatial.neighbor"
spatial.solve	Solve S b = x
spatial.subset	Subset an Object of Class "spatial.neighbor"
spatial.sum	Sum of Two Objects of Class "spatial.neighbor"
spatial.weights	Compute a Spatial Weight Matrix
spher.cov	Theoretical Distance Based Covariance Functions
spher.vgram	Theoretical Variogram Functions
spp	Spatial Point Pattern Objects
summary.hexbin	Summary Method for a Hexagonally Binned Object
summary.slm	Summary Method for Spatial Linear Models
summary.spp	Summary Method for a Spatial Point Pattern Object

Spatial Statistics Module, (cont.)

triangulate	Delaunay's Triangulation
twoway.formula	Fit of a Two-Way Table (Formula Method)
variogram	Empirical Variogram
variogram.cloud	Calculate Variogram Cloud
wheat	Wheat Grain and Straw Yield
write.geoeas	Write A GEO-EAS Data File
xy2cell	Compute Hexagon Cell Ids From x and y

Statistical Inference

lrt	Computes a Likelihood Ratio Test
lrt.slm	Likelihood Ratio Test for Spatial Linear Models

Appendix C

Sample Data Sets

The data sets described in this section are included with S+SPATIALSTATS. The information in this section is also found in the on-line help. To see how to access the on-line help, look at section 2.2.

`aquifer`	Wolfcamp Aquifer Data	`aquifer`

SUMMARY

    The `aquifer` data frame is a spatial data set with 85 rows and 3 columns: `easting`, `northing` and `head`. The data are from the Wolfcamp Aquifer in West Texas/New Mexico, U.S.A.

DATA DESCRIPTION

    This data frame contains the following columns:

  `easting` relative longitude position.

`northing` relative latitude position.

    `head` piezometric-head in feet above sea level.

SOURCE

    Cressie (1989) lists the data and uses it to illustrate kriging. The original data are from Harper and Furr (1986).

REFERENCES

    Cressie, Noel. (1989). Geostatistics. *The American Statistician, 43*, 197-202.

    Harper, W. V. and Furr, J. M. (1986). Geostatistical Analysis of Potentiometric Data in the Wolfcamp Aquifer of the Palo Duro Basin, Texas. Technical Report ONWI-587, Battelle Memorial Institute, Columbus, OH.

`bramble`	Bramble Cane Data	`bramble`

SUMMARY

    The `bramble` data frame is a spatial point process data set of the locations of 359 newly emerged bramble canes.

DATA DESCRIPTION

    This data frame contains the following columns:

  `x` relative x coordinate.

  `y` relative y coordinate.

SOURCE

    Hutchings (1979) describes the data from a 9 meter square plot "staked out within a dense thicket of bramble". The values were copied from Diggle (1983). The locations of one and two year old canes from the same plot were also recorded, they are not included here.

REFERENCES

    Diggle, Peter. (1983). *Statistical Analysis of Spatial Point Patterns.* Academic Press, New York.

    Hutchings, M. J. (1979). Standing crop and pattern in pure stands of *Mercurialis perennis* and *Rubus fruticosus* in mixed deciduous woodland. *Oikos* **31**, 351-357.

| `coal.ash` | Coal Ash Data | `coal.ash` |

SUMMARY

The `coal.ash` data frame is a spatial data set with 208 rows and 3 columns: `x`, `y` and `coal`.

DATA DESCRIPTION

The data frame contains the following columns:

`x` relative x position.

`y` relative y position.

`coal` percent coal ash at location `x`, `y`.

The original data were obtained from Gomez and Hazen (1970, Tables 19 and 20) for the Robena Mine Property in Greene County, Pennsylvania, USA. The data come from the Pittsburgh coal seam which is associated with a deltaic sedimentation system that includes much of southwestern Pennsylvania, northeastern Ohio and northern West Virginia. Cressie (1993) has reoriented the locations so that it appears to run in an east-west and north-south direction.

SOURCE

From Cressie (1993) who uses the data to illustrate kriging.

REFERENCES

Cressie, Noel. (1993). *Statistics For Spatial Data,* Revised Edition. Wiley, New York.

Gomez, M. and Hazen, K. (1970). Evaluating sulfur and ash distribution in coal seams by statistical response surface regression analysis. U.S. Bureau of Mines Report RI 7377.

| `iron.ore` | Iron Ore Data | `iron.ore` |

SUMMARY

The `iron.ore` data frame is a spatial data set with 112 rows and 4 columns: `easting`, `northing`, `ore`, and `residuals`.

DATA DESCRIPTION

The data frame contains the following columns:

`easting` relative longitude position.

`northing` relative latitude position.

`ore` percent iron (Fe_2O_3) at location `easting`, `northing`.

`residuals` residuals from a median polish fit to the data. These are analyzed in Cressie (1991) and Zimmerman and Zimmerman (1991).

The original data have been modified slightly (for confidentiality reasons) by multiplying by and adding unspecified constants. Grid spacing is 50 meters by 50 meters.

SOURCE

The modified data appear in Cressie (1986) and also in Zimmerman and Zimmerman (1991).

REFERENCES

Cressie, Noel. (1986). Kriging nonstationary data. *Journal American Statistical Association,* **81**, 625-634.

Zimmerman, Dale L. and Zimmerman, M. Bridget. (1991). A comparison of spatial semivariogram estimators and corresponding ordinary kriging predictors. *Technometrics,* **33**, 77-91.

lansing	Lansing Woods Tree Data	lansing

SUMMARY

The `lansing` data frame is a marked spatial point process data set of 1217 trees. The location and species type is recorded for each tree.

DATA DESCRIPTION

This data frame contains the following columns:

x   relative x coordinate.

y   relative y coordinate.

species   a factor identifying the species, either `"hickory"` or `"maple"`. The data set from Diggle (1983) also included 346 red oaks, 448 white oaks, 135 black oaks and 105 miscellaneous trees, these trees are not included in this data set.

SOURCE

The data are from a 19.6 acre square plot in Lansing Woods, Clinton County, Michigan, USA. It was originally described by Gerrard (1969). The values were copied from Diggle (1983).

REFERENCES

Diggle, Peter. (1983). *Statistical Analysis of Spatial Point Patterns.* Academic Press, New York.

Gerrard, D. J. (1969). Competition quotient: a new measure of the competition affecting individual forest trees. Research Bulletin No. 20, Agricultural Experiment Station, Michigan State University.

quakes.bay	Bay Area Earthquakes	quakes.bay

SUMMARY

The `quakes.bay` data frame is a marked spatial point pattern data set listing the location, time and magnitude of earthquakes in the San Francisco Bay Area from 1962 to 1981.

DATA DESCRIPTION

The data frame contains the following columns:

longitude   negative longitude of quake location.

latitude   latitude of quake location.

magnitude   the size of the quake.

date   the date of the quake, an object of class `dates`.

hour   the hour of the quake.

minute   the minute of the quake.

second   the second of the quake.

SOURCE

The data are from Peter Guttorp, Statistics Department, University of Washington, Seattle, Washington.

| `quakes.wash` | Washington State Earthquakes In 1980 | `quakes.wash` |

SUMMARY

The `quakes.wash` data frame is a marked spatial point pattern data set listing the location, time, and magnitude of earthquakes in and around Washington State in 1980. The Mount St Helens volcano in Washington State erupted several times in 1980 including the massive eruption on May 18.

DATA DESCRIPTION

The data frame contains the following columns:

`longitude` negative longitude of quake location.

`latitude` latitude of quake location.

`magnitude` the size of the quake.

`date` the date of the quake, an object of class `dates`.

`hour` the hour of the quake.

`minute` the minute of the quake.

`second` the second of the quake.

SOURCE

The data are from Peter Guttorp, Statistics Department, University of Washington, Seattle, Washington.

| `scallops` | Scallop Abundance Data | `scallops` |

SUMMARY

The `scallops` data frame is a spatial data set listing the catch of scallops from a 1990 National Marine Fisheries Service trawl survey in the Atlantic Ocean. The survey area runs from the Delmarva Peninsula off the coast of Virginia and Maryland up to the George Banks.

DATA DESCRIPTION

This data frame contains the following columns:

`strata` a factor indicating the National Marine Fisheries Service (NMFS) 4 digit strata designator in which the sample was taken.

`sample` sample number per year ranging from 1 to approximately 450.

`lat` location in terms of latitude of each sample in the Atlantic Ocean.

`long` location in terms of negative longitude of each sample in the Atlantic Ocean.

`tcatch` total number of scallops caught at the ith sample location. This is `prerec + recruits`.

`prerec` number of scallops whose shell length is smaller than 70 millimeters.

`recruits` number of scallops whose shell length is 70 millimeters or larger.

SOURCE

From Ecker and Heltshe (1994) who present a geostatistical analysis of the data.

REFERENCES

Ecker, Mark D. and Heltshe, James F. (1994) Geostatistical estimates of scallop abundance. In *Case Studies in Biometry*. Nicholas Lange, Louise Ryan, Lynne Billard, David Brillinger, Loveday Conquest, and Joel Greenhouse, eds. New York: Wiley, pp. 107-124.

| sids | Sudden Infant Death Syndrome Data 1974-1978 | sids |

SUMMARY

The `sids` data frame contains counts of sudden infant deaths for 1974 to 1978 along with related information in 100 counties in North Carolina.

DATA DESCRIPTION

The data frame contains the following columns:

id an identifying number for the county (in the range 1 to 100).

easting the x coordinate of the county relative to an arbitrary origin, where the x-axis is parallel to the latitude.

northing the y coordinate of the county relative to an arbitrary origin, where the y-axis is parallel to the longitude.

sid the number of SIDS deaths in the county during the years 1974 to 1978.

births the number of births in the county during years 1974 to 1978.

nwbirths the number of nonwhite births, 1974 to 1978.

group a grouping of the counties to one of 12 parcels of contiguous counties.

sid.ft a dependent variable obtained as the Freeman-Tukey transformation of the number of SIDS cases to the number of births during the years 1974 to 1978.

nwbirths.ft an independent variable derived as the Freeman-Tukey square root transformation of the ratio nonwhite births to the number of births during the years 1974 to 1978.

SOURCE

The data are listed and analyzed in Cressie and Chan (1989).

REFERENCES

Cressie, Noel and Chan, Ngai H. (1989). Spatial modeling of regional variables. *Journal of the American Statistical Association* **84**, 393-401.

SEE ALSO

sids2, sids.neighbor

| sids2 | Sudden Infant Death Syndrome Data 1979-1984 | sids2 |

SUMMARY

The `sids2` data frame contains counts of sudden infant deaths for 1979 to 1984 along with related information in 100 counties in North Carolina.

DATA DESCRIPTION

The data frame contains the following columns:

id an identifying number for the county (in the range 1 to 100).

easting the x coordinate of the county relative to an arbitrary origin, where the x-axis is parallel to the latitude.

northing the y coordinate of the county relative to an arbitrary origin, where the y-axis is parallel to the longitude.

sid the number of SIDS deaths in the county during the years 1979 to 1984.

births the number of births in the county during years 1979 to 1984.

nwbirths the number of nonwhite births, 1979 to 1984.

group a grouping of the counties to one of 12 parcels of contiguous counties.

SOURCE

The data are listed and analyzed in Cressie and Chan (1989).

REFERENCES

Cressie, Noel and Chan, Ngai H. (1989). Spatial modeling of regional variables. *Journal of the American Statistical Association* **84**, 393-401.

SEE ALSO

`sids`, `sids.neighbor`.

`sids.neighbor`	Neighbors for Sudden Infant Death Syndrome Data	`sids.neighbor`

SUMMARY

An object of class `"spatial.neighbor"` containing the neighbor specification for the sudden infant deaths data (`sids`).

DATA DESCRIPTION

Three hundred and ninety four neighbor relationships are specified. See `spatial.neighbor.object` for a description of the data within an object of class `"spatial.neighbor"`.

SOURCE

The data are listed and analyzed in Cressie and Chan (1989).

REFERENCES

Cressie, Noel and Chan, Ngai H. (1989). Spatial modeling of regional variables. *Journal of the American Statistical Association* **84**, 393-401.

SEE ALSO

`sids`, `sids2`, `spatial.neighbor.object`.

`wheat`	Wheat Grain and Straw Yield	`wheat`

SUMMARY

The `wheat` data frame is a spatial data set listing yields of grain and straw from 500 plots of wheat laid out in 25 x 20 lattice covering approximately 1 acre in total area. The data were collected in 1910 at the Rothamsted Experimental Station in England (Mercer and Hall, 1911).

DATA DESCRIPTION

The data frame contains the following columns:

`x` the relative x location of the plot. The range of `x` is 1 to 25.
`y` the relative y location of the plot. The range of `y` is 1 to 20.
`grain` the yield of grain for the plot in pounds.
`straw` the yield of straw for the plot in pounds.

SOURCE

The data are from Mercer and Hall (1911).

REFERENCES

Mercer, W. B. and Hall, A. D. (1911). The experimental error of field trials. *Journal Agricultural Science (Cambridge),* **4**, 107-132.

Appendix D

Function Reference

The functions described in this section are included with S+SPATIALSTATS. The information in this section is also found in the on-line help. To see how to access the on-line help, look at section 2.2.

| `anisotropy.plot` | Explore Corrections For Geometric Anisotropy | `anisotropy.plot` |

DESCRIPTION

Computes corrections for geometric anisotropy for two dimensional spatial data and plots variograms based on the corrections.

USAGE

```
anisotropy.plot(formula=formula(data), data=sys.parent(), subset,
                na.action, lag=<<see below>>, nlag=20,
                tol.lag=lag/2, maxdist=<<see below>>,
                angle=seq(0, 180, length = 5),
                ratio=seq(1.25, 2, length = 4),
                minpairs=6, method="classical",
                smooth=T, plot.it=T, panel=panel.xyplot, ...)
```

REQUIRED ARGUMENTS

formula   formula defining the response and the predictors. In general, its form is:

```
z ~ x + y
```

The z variable is a numeric response. Variables x and y are the locations. All variables in the formula must be vectors of equal length with no missing values (NAs). The formula may also contain expressions for the variables, e.g. `sqrt(count)`, `log(age+1)` or `I(2*x)`. (The `I()` is required since the `*` operator has a special meaning on the right side of a formula.

OPTIONAL ARGUMENTS

data   an optional data frame in which to find the objects mentioned in `formula`.

subset   expression saying which rows of the data should be used in the fit. This can be a logical vector (which is replicated to have length equal to the number of observations), or a numeric vector indicating which observation numbers are to be included, or a character vector of the row names to be included.

na.action   a function to filter missing data. This is applied to the `model.frame` after any `subset` argument has been used. The default (with `na.fail`) is to create an error if any missing values are found. A possible alternative is `na.omit`, which deletes observations that contain one or more missing values.

lag   a numeric value, the width of the lags. If missing, `lag` is set to `maxdist / nlag`.

nlag   an integer, the maximum number of lags to calculate.

tol.lag   a numeric value, the distance tolerance.

maxdist   the maximum distance to include in the returned output. The default is half the maximum distance in the transformed data.

angle   a vector of direction angles (in degrees, clockwise from North) to consider as directions of anisotropy.

ratio   a vector of ratios of anisotropy. These should all be greater than 1.

minpairs   the minimum number of pairs of points (minimum value for np) that must be used in calculating a variogram value. If np is less than `minpairs` then that value is dropped from the variogram.

method   a character string to select the method for estimating the variogram. The possible values are `"classical"` for Matheron's (1963) estimate and `"robust"` for Cressie and Hawkins (1980) robust estimator. Only the first character of the string needs to be given.

smooth   a logical flag, if TRUE, a loess smooth line will be drawn for each variogram panel. If `panel` is supplied then this value is ignored.

panel   a panel function to be used in plotting the variograms. If `plot.it=FALSE` then this value is ignored.

plot.it   a logical flag, if TRUE, a plot of all the variogram will be drawn.

...   additional arguments to be passed down to the panel function for plotting.

VALUE

a data frame with columns:

distance the average distance for pairs in the lag.
   gamma the variogram estimate.
      np the number of pairs in each lag.
   angle a factor denoting the angle for the geometric anisotropy.
   ratio a factor with levels denoting the ratio for the geometric anisotropy.

SIDE EFFECTS

If `plot.it=TRUE` (the default) the variogram for each combination of `angle` and `ratio` is plotted. The plot is drawn using `xyplot`.

DETAILS

For each combination of `angle` and `ratio` the locations are corrected for geometric anisotropy. The correction consists of multipling each location pair $(x[i],y[i])$ by the symmetric 2 x 2 matrix `A` where `A[1,1]=cos(angle)^2+ratio*sin(angle)^2`, `A[1,2]=(1- ratio) * sin(angle) * cos(angle)` and `A[2,2]=sin(angle)^2+ratio*cos(angle)^2`. See Journel and Huijbregts (1978, pp 179-181). The variogram is then estimated using these corrected locations.

REFERENCES

Cressie, N. and Hawkins, D. M. (1980). Robust estimation of the variogram. *Mathematical Geology* **12**, 115-125.

Journel, A. G. and Huijbregts, Ch. J. (1978). *Mining Geostatistics*. Academic Press, New York.

Matheron, G. (1963). Principles of geostatistics. *Economic Geology* **58**, 1246-1266.

SEE ALSO

`loc`, `variogram`, `xyplot`.

EXAMPLES

```
anisotropy.plot(log(tcatch+1) ~ long + lat, data=scallops, lag=.075)
```

---

bbox	Bounding Box for a Spatial Point Pattern	**bbox**

DESCRIPTION

Generates a bounding box for a list of points in particular for an object of class `"spp"`.

USAGE

```
bbox(x, y)
```

REQUIRED ARGUMENTS

x an object of class `"spp"`, a spatial point pattern object. It could also be a list with named components `"x"` and `"y"` or a vector of x-coordinates when argument `y` is given.

OPTIONAL ARGUMENTS

y numeric vector. Needed if `x` is a vector in which case it must have the same length as `x`.

VALUE

a list with two named components `"x"` and `"y"` containing the coordinates of the corners of the bounding box of the data in `x` (and `y`, if given).

EXAMPLES

```
maples <- spp(lansing[lansing$species == "maple",], drop=T)
bbox(maples)
```

`boxplot.vgram.cloud`	Boxplots of a Variogram Cloud Object	`boxplot.vgram.cloud`

DESCRIPTION

Produces side by side boxplots from a variogram cloud object.

USAGE

```
boxplot.vgram.cloud(x, nint=20, group, style.bxp="att",
                    plot=T, ...)
```

REQUIRED ARGUMENTS

x   an object of class `"vgram.cloud"`.

OPTIONAL ARGUMENTS

nint   an integer specifying the number of equal width intervals of `x$distance` to split up the `x$gamma` values. This is ignored if `group` is supplied.

group   a vector of groups to split `x$gamma` into. There will be one boxplot drawn for each unique value of `group`. This must be the same length as `x$gamma`. If supplied, the value of `nint` is ignored.

style.bxp   the style for the boxplots. See the `boxplot` help file for possible values. The default value of `att` draws better boxplots for a variogram cloud than the style used in the default `boxplot` function.

plot   if `TRUE`, the boxplot will be produced; otherwise, the calculated summaries of the arguments are invisibly returned.

...   other arguments to the default boxplot function can be supplied.

Graphical parameters may also be supplied as arguments to this function (see `par`). In addition, the high-level graphics arguments described under `par` and the arguments to `title` may be supplied to this function. However, `boxplot` will always use linear axes: the `log` and `[xy]axt` arguments are ignored.

VALUE

if `plot` is `TRUE`, the function `bxp` is invoked with these components, plus optional `width`, `varwidth`, `notch`, and `style` (and associated parameters), to produce the plot. Note that `bxp` returns a vector of box centers.

if `plot` is `FALSE`, an invisible list with the components listed below:

stats   matrix (of size 5 by the number of boxes) giving the upper extreme (excluding outliers), upper quartile, median, lower quartile, and lower extreme (excluding outliers) for each box. By default, anything farther than 1.5 times the Inter-Quartile Range is considered an outlier. See the Details section below and the `range` argument above.

n   the number of observations in each group.

conf   matrix (of size 2 by the number of boxes) giving approximate 95% confidence limits for the median. The limits are functions of the quartiles, so a few outliers have little effect on them.

out   optional vector of outlying points (outliers). See the Details section below.

group   vector giving the box to which each point in `out` belongs.

names   names for each box.

SIDE EFFECTS

If `plot` is `TRUE`, a plot is created on the current graphics device.

DETAILS

This function is a method for the generic function `boxplot` for class `vgram.cloud` . It can be invoked by calling `boxplot` for an object of the appropriate class, or directly by calling box-

`plot.vgram.cloud` regardless of the class of the object.

By default, whiskers are drawn to the nearest value not beyond a standard span from the quartiles; points beyond (outliers) are drawn individually. Giving `range=0` forces whiskers to the full data range. Any positive value of `range` multiplies the standard span by this amount. The standard span is 1.5*(Inter-Quartile Range).

REFERENCES

Cressie, Noel. (1993). *Statistics For Spatial Data,* Revised Edition. Wiley, New York.

SEE ALSO

`boxplot, plot.vgram.cloud, variogram.cloud`.

EXAMPLES

```
scallop.vgcld <- variogram.cloud(log(tcatch+1) ~ loc(long,lat),
    data=scallops)
boxplot(scallop.vgcld)
```

CAR	Conditional Spatial Autoregression Object	**CAR**

DESCRIPTION

An object of class `"cov.family"` containing functions for fitting conditional autoregression models when used as input to the spatial linear models function, `slm`. See `cov.family.object` for a discussion of the attributes contained in the CAR object. The discussion here centers around the covariance matrix model which the CAR object supports.

DETAILS

Let N denote a neighbor matrix obtained from an object of class `"spatial.neighbor"`, and let Σ denote the covariance matrix of a vector y of spatially correlated dependent variables. Finally, let W denote a diagonal matrix of weights. Then a conditional autoregression model assumes that

$$\Sigma = (I - \rho N)^{-1} \, W \, \sigma^2$$

where ρ and σ are scale parameters to be estimated. This model for the covariance matrix can be generalized to multiple matrices N_i using multiple parameters ρ_i as follows:

$$\Sigma = \left[\sum_i (I - \rho_i N_i) \right]^{-1} W \, \sigma^2,$$

where the N_i are specified through component `matrix` of the `"spatial.neighbor"` object (see routine `spatial.neighbor`).

The "regression" aspect of a spatial regression fits the multivariate normal mean vector

$$\mu = E(y|x) = x\beta$$

for unknown parameters β. The multivariate normal likelihood is expressed in terms of the unknowns ρ, σ, and β. The CAR object assumes that a profile likelihood for ρ is fit.

The CAR model can be expressed as an autoregressive model for the spatial parameters as follows:

$$y = X\beta + \rho N(y - X\beta) + W^{1/2}\varepsilon.$$

For the CAR model, the residuals, ε, defined here are not independent. The formula for y allows one to decompose the sum of squares in y into three components (see Haining, 1990, page 258): 1) the trend, $X\beta$; 2) the noise,

$$W^{1/2}\varepsilon = (I - \rho N)(y - X\beta);$$

and 3) the signal,

$$y - X\beta - W^{1/2}\varepsilon.$$

Function `residual.fun` of the `CAR` object computes ε, the standardized residuals, and routine `slm` returns these in component `residuals`. The estimated trend, $X\beta$, is returned by routine `slm` as the fitted values.

Two functions are required to compute the profile likelihood: 1) a function for computing the determinant $|\Sigma|$, and 2) a function for computing the vector $\Sigma^{-1}z$ for arbitrary vector z. When the single neighbor matrix N is symmetric, the determinant can be expressed and efficiently computed as a function of the eigenvalues of N. If N is not symmetric, or if the dimension of N is large (over 150), then sparse matrix routines by Kundert (1988) are used to compute the determinant of Σ. Because the covariance matrix is parameterized in terms of its inverse, the computation of $\Sigma^{-1}z$ is simple and is carried out using (sparse) matrix multiplication. See routine `spatial.multiply`.

REFERENCES

Haining, R. (1990). *Spatial Data Analysis in the Social and Environmental Sciences.* Cambridge University Press. Cambridge.

Kundert, Kenneth S. and Sangiovanni-Vincentelli, Alberto (1988). A Sparse Linear Equation Solver. Department of EE and CS, University of California, Berkeley.

SEE ALSO

`cov.family.object`, `slm`, `slm.fit`, `slm.nlminb`, `spatial.neighbor`, `MA`, `SAR`, `spatial.multiply`.

---

cell2xy	Compute x,y Coordinates From Hexagon Cell Ids	**cell2xy**

DESCRIPTION

Computes x and y coordinates from hexagonal cell indices.

USAGE

`cell2xy(bin)`

REQUIRED ARGUMENTS

`bin` an object of class `"hexbin"`.

VALUE

a data frame with components x and y containing the cartesian coordinates corresponding to the centers of the hexagonal bins described in `bin`.

DETAILS

The `bin` object contains all the information that is usually needed. This function reduces storage by allowing the user to call it whenever the lattice center coordinates are needed. To plot the hexagonal centers use `plot(as.list(cell2xy(bin)))`.

REFERENCES

Carr, D. B., Olsen, A. R. and White, D. (1992). Hexagon mosaic maps for display of univariate and bivariate geographical data. *Cartography and Geographics Information Systems,* **19**, 228-236.

SEE ALSO

`summary.hexbin, plot.hexbin, identify.hexbin, smooth.hexbin, erode.hexbin, xy2cell.`

EXAMPLES

```
bin1 <- hexbin(x,y)
lattice <- cell2xy(bin1)
plot(as.list(lattice))
hexagons(bin1,density=0,border=T)
```

`check.islands`	Detect Isolated Spatial Regions	`check.islands`

DESCRIPTION

Given an object of class `"spatial.neighbor"` detects spatial units that have no neighbors (islands).

USAGE

`check.islands(x, remap=F)`

REQUIRED ARGUMENTS

x an object of class `"spatial.neighbor"`.

OPTIONAL ARGUMENTS

remap logical flag: if there is an island, should we recode the indexing of the spatial contiguity matrix to eliminate the rows and columns with all zeroes? That is, should we renumber components `row.id` and `col.id` of the spatial neighbor object?

VALUE

if `remap=FALSE` the list of existing islands is returned. Otherwise, an object of class `"spatial.neighbor"` with remapped `row.id` and `col.id`.

SIDE EFFECTS

the attribute `"nregion"` of x may be recomputed when `remap=T`.

SEE ALSO

`spatial.neighbor, spatial.subset, spatial.weights`

EXAMPLES

`sids.nhbr2 <- check.islands(sids.neighbor,remap=T)`

`correlogram`	Empirical Correlogram and Covariogram	`correlogram`

DESCRIPTION

Computes the empirical correlogram or covariogram for two dimensional spatial data. Multiple correlograms and covariograms for different directions can be computed.

USAGE

```
correlogram(formula, data=<<see below>>, subset=<<see below>>,
        na.action=<<see below>>, lag=<<see below>>, nlag=20,
        tol.lag=lag/2, azimuth=0, tol.azimuth=90, bandwidth=1e21,
        maxdist=<<see below>>, minpairs=6)
covariogram(formula, data=<<see below>>, subset=<<see below>>,
        na.action=<<see below>>, lag=<<see below>>, nlag=20,
        tol.lag=lag/2, azimuth=0, tol.azimuth=90, bandwidth=1e21,
        maxdist=<<see below>>, minpairs=6)
```

REQUIRED ARGUMENTS

formula   formula defining the response and the predictors. In general, its form is:

```
z ~ x + y
```

The z variable is a numeric response. Variables x and y are the locations. All variables in the formula must be vectors of equal length. The formula may also contain expressions for the variables, e.g. `sqrt(count)`, `log(age+1)` or `I(2*x)`. (The `I()` is required since the * operator has a special meaning on the right side of a formula. The right hand side may also be a call to the `loc` function e.g. `loc(x,y)`. The `loc` function can be used to correct for geometric anisotropy, see the `loc` help file.

OPTIONAL ARGUMENTS

data   an optional data frame in which to find the objects mentioned in `formula`.

subset   expression saying which subset of the rows of the data should be used in the fit. This can be a logical vector (which is replicated to have length equal to the number of observations), or a numeric vector indicating which observation numbers are to be included, or a character vector of the row names to be included.

na.action   a function to filter missing data. This is applied to the `model.frame` after any `subset` argument has been used. The default (with `na.fail`) is to create an error if any missing values are found. A possible alternative is `na.omit`, which deletes observations that contain one or more missing values.

lag   a numeric value, the width of the lags. If missing, `lag` is set to `maxdist / nlag`.

nlag   an integer, the maximum number of lags to calculate.

tol.lag   a numeric value, the distance tolerance.

azimuth   a vector of direction angles in degrees, measured from North-South. A separate correlogram will be estimated for each direction.

tol.azimuth   angle tolerance in degrees. A `tol.azimuth` of 90 or greater (the default) results in an omnidirectional correlogram.

bandwidth   the maximum bandwidth, the deviation from the direction orthogonal to the direction angle.

maxdist   the maximum distance to include in the returned output. The default is half the maximum distance in the data.

minpairs   the minimum number of pairs of points (minimum value for np) that must be used in calculating a correlogram or covariogram value. If np is less than `minpairs` then that value is dropped from the result.

VALUE

an object of class `"correlogram"` or `"covariogram"` that inherits from `"variogram"` and `"data.frame"` with columns:

distance the average distance for pairs in the lag.
      rho the correlogram estimate (if `correlogram` was called).
      cov the covariogram estimate (if `covariogram` was called).
      np the number of pairs in each lag.
 azimuth a factor denoting the angular direction.

The return object has an attribute `call` with an image of the call that produced the object.

DETAILS

There are plot methods for classes `"correlogram"` and `"covariogram"`. The `print` and `summary` methods for class `"variogram"` can be used through inheritance.

The covariogram is a measure of spatial covariance as a function of distance. The correlogram is a standardized covariogram where the values are between -1 and 1. These functions make a call to the function `variogram` with the argument `type` set to `"correlogram"` or `"covariogram"`. The computations are based on a modified version of the gamv2 subroutine from GSLIB (Deutsch and Journel, 1992).

REFERENCES

Cressie, Noel A. C. (1993). *Statistics for Spatial Data,* Revised Edition. Wiley, New York.

Deutsch, Clayton V. and Journel, Andre G. (1992). *GSLIB Geostatistical Software Library and User's Guide.* Oxford University Press, New York.

SEE ALSO

`loc`, `plot.variogram`, `variogram`.

EXAMPLES

```
# an omnidirectional correlogram
c1 <- correlogram(log(tcatch+1) ~ lat + long, data=scallops)
plot(c1)
# correlograms in 0, 45, 90 and 135 degrees directions
c2 <- correlogram(log(tcatch+1) ~ loc(lat,long), data=scallops,
            azimuth=c(0,45,90,135), tol.azimuth=22.5)
plot(c2)
```

---

cov.family.object	Spatial Family Object	**cov.family.object**

DESCRIPTION

Class of objects used for defining the covariance structure in a spatial regression. An object of class `"cov.family"` contains information required to optimize a profile likelihood in a spatial regression analysis, and is required input for `slm`.

GENERATION

No functions for constructing objects of class `"cov.family"` exist. Objects of the class have already been created for the standard covariance structures, CAR, SAR, and MA. If you choose to construct a new `"cov.family"` object, the easiest way to proceed is to copy an existing `"cov.family"` object, and modify the new object.

METHODS

The class `"cov.family"` has no associated methods.

INHERITANCE

Class "cov.family" does not inherit from any other class.

STRUCTURE

The "cov.family" object is implemented as a list with the components given below.

name the name of the covariance family.

det.fun a function for computing the determinant of the estimated spatial covariance matrix.

```
det.fun(parameters, spatial.arglist, initial)
```

ARGUMENTS

parameters - a vector containing the current values of the covariance matrix model parameters. A profile likelihood for the covariance matrix model parameters is optimized given optimal values for the parameters in the linear model.

spatial.arglist - a list of additional arguments to be used in the spatial model of the covariance matrix. See routine slm.

initial - a list containing the return value from the initialize function described below.

VALUE

the determinant of the current spatial covariance matrix.

solve.fun a function to solve for y in the linear equation x = S y, where S is a square symmetric covariance matrix, and x is a known matrix. The solution of this equation defines the spatial covariance matrix S.

```
solve.fun(parameters, x, spatial.arglist, initial)
```

ARGUMENTS

parameters - a vector containing the current values of the spatial parameters required by the model for the covariance matrix (see above).

x - a matrix containing the right hand sides.

spatial.arglist - a list of additional arguments to be used in defining the spatial model for the covariance matrix. See routine slm.

initial - a list containing the return value from the initialize function described below. These may be used as additional variables when computing the spatial covariance matrix.

VALUE

the matrix of solutions x to the linear equations y = S x, where S is the spatial covariance matrix specified by this family of covariance matrices.

initial.parameters a function to compute or set initial parameter estimates.

```
initial.parameters(spatial.arglist)
```

ARGUMENTS

spatial.arglist - a list of additional arguments to be used in the spatial model of the covariance matrix. See routine slm.

VALUE

a vector containing the initial parameter estimates.

`initialize` a function used to compute initial statistics for use in the iterative algorithm.

> `initialize(spatial.arglist, subset)`

ARGUMENTS
`spatial.arglist` - a list of additional arguments to be used in the spatial model of the covariance matrix. See routine `slm`.

`subset` - a logical vector (which is replicated to have length equal to the number of observations), a numeric vector, or a character vector of the row names. This vector is used to indicate which observation numbers are to be included in the analysis. All observations are included by default. See routine `slm`.

VALUE
a list to be used by the `det.fun` and the `solve.fun` to simplify and speed the computations. For many covariance models, this is the eigenvalues of the symmetric spatial neighbor matrix.

`finalize` a function called at the termination of the computations for the `cov.family` to allow cleanup operations, should these be necessary.

> `finalize(spatial.arglist, initial)`

ARGUMENTS
`spatial.arglist` - a list of additional arguments to be used in the spatial model of the covariance matrix. See routine `slm`.

`initial` - a list containing the values returned by the `cov.family$initialize` function. This list depends upon the `cov.family`.

`residual.fun` a function to compute the residuals for the spatial model.

> `residual.fun(x, y, beta, parameters, spatial.arglist)`

ARGUMENTS
`x` - the independent variable(s) (if any).

`y` - the dependent variable.

beta the estimated linear model coefficients.

parameters - the estimated covariance model parameters.

`spatial.arglist` - a list of additional arguments to be used in the spatial model of the covariance matrix. See routine `slm`.

VALUE
The return value contains an estimate of the residuals (standardized to be independent with constant variance).

DETAILS
The six functions, `det.fun`, `solve.fun`, `residual.fun`, `initial.parameters`, `initialize`, and `finalize` are used by the spatial regression routine to define a spatial regression. Currently three models for a spatial regression can be fit: CAR, SAR, and MA.

SEE ALSO

CAR, SAR, MA, slm.

---

erode.hexbin	Erode a Hexagonally Binned Image	erode.hexbin

DESCRIPTION

Algorithm to remove counts from hexagonal cells at a rate proportional to the cell's exposed surface area. When a cell becomes empty, the cell is removed and its removal order noted. Cell removal increases the the exposure of neighboring cells. The last cell removed is a type of bivariate median.

USAGE

erode.hexbin(bin, cdfcut=0.5)

REQUIRED ARGUMENTS

bin an object of class "hexbin".

OPTIONAL ARGUMENTS

cdfcut numeric value giving the fraction of the total counts to be extracted. All cells will be extracted if cdfcut=0.

VALUE

an object of class "hexbin" with high count cells and a component named erode which gives the erosion order.

DETAILS

The algorithm performs gray-level erosion on the extracted cells. Each erosion cycle removes counts from cells. The counts removed for each cell are a multiple of the cell's exposed-face count. The algorithm chooses the multiple so that at least one cell will be empty or have a count deficit on each erosion cycle.

The erode component of the resulting object (a data frame) contains an erosion number for each cell. The value of erode is six times the erosion cycle at cell removal minus the cell deficit at removal.

Cells with low values are eroded first. The cell with the highest erosion number is a candidate bivariate median. A few ties in erode are common.

REFERENCES

Carr, D. B. (1991). Looking at large data sets using binned data plots. In *Computing and Graphics in Statistics*. A. Buja and P. Tukey, eds. Springer-Verlag, New York. pp. 7-39.

SEE ALSO

smooth.hexbin, hexbin, plot.hexbin.

EXAMPLES

```
bin <- hexbin(rnorm(500), rnorm(500))
smbin  <- smooth.hexbin(bin)
erobin <- erode.hexbin(smbin)
screenpar <- plot(erobin, style="lat", minarea=1, maxarea=1,
        density=0, border=T)  # Show erosion order
oldpar <- par(screenpar)  # Reset graphical parameters
xy <- cell2xy(erobin)
text(xy$x, xy$y, as.character(erobin$erode))
par(oldpar)  # Restore old graphical parameters
```

exp.cov	Theoretical Distance Based Covariance Functions	**exp.cov**

DESCRIPTION

Computes theoretical covariance function at supplied distance values. Models include exponential, spherical, and gaussian covariances.

USAGE

```
exp.cov(distance, range, sill=1, nugget=0, eps=1.0e-7)
spher.cov(distance, range, sill=1, nugget=0, eps=1.0e-7)
gauss.cov(distance, range, sill=1, nugget=0, eps=1.0e-7)
```

REQUIRED ARGUMENTS

distance a vector of distances to compute the covariance for.

range the range value.

OPTIONAL ARGUMENTS

sill the sill value. This is the absolute sill, the variance (covariance at distance = 0) is sill + nugget.

nugget the nugget effect.

eps any distance less than eps will be set to nugget + sill.

VALUE

a vector of covariance values at the supplied distances.

DETAILS

These functions are used with the krige function.

SEE ALSO

exp.vgram, krige

EXAMPLES

```
dist <- seq(0,5,length=50)
plot(dist,exp.cov(dist,range=2,nugget=.2),type='l')
```

exp.vgram	Theoretical Variogram Functions	**exp.vgram**

DESCRIPTION

Computes theoretical isotropic variograms at supplied distance values. Models include exponential, spherical, gaussian, linear and power variograms.

USAGE

```
exp.vgram(distance, range, sill=1, nugget=0)
spher.vgram(distance, range, sill=1, nugget=0)
gauss.vgram(distance, range, sill=1, nugget=0)
linear.vgram(distance, slope, nugget=0)
power.vgram(distance, slope, range=1, nugget=0)
```

REQUIRED ARGUMENTS

distance a vector of distances to compute the variogram for.

range the range value for the exponential, spherical and gaussian variograms or the exponent value for the power variogram.

slope the slope for the variogram, only appropriate for `linear.vgram` and `power.vgram`.

OPTIONAL ARGUMENTS

sill the sill value for the variogram, only applies to the exponential, spherical and gaussian models. This is the absolute sill, it is the maximum value of the variogram minus any nugget effect.

nugget the nugget effect.

VALUE

a vector of variogram values at the supplied distances.

SEE ALSO

`exp.cov`, `variogram`.

EXAMPLES

```
dist <- seq(0,6,length=100)
plot(dist,spher.vgram(dist,range=3,nugget=.2),type='l')
```

Fhat	EDF of Origin-to-Point Nearest Neighbor Distances	Fhat

DESCRIPTION

Computes and plots the empirical distribution function (EDF) of the origin-to-point nearest neighbor distances for a spatial point pattern.

USAGE

```
Fhat(obj1, obj2, nx=sqrt(n), ny=sqrt(n), dist.fhat=all.dists,
     plot.it=T)
```

REQUIRED ARGUMENTS

obj1 an object of class `"spp"` representing a spatial point pattern, or a data frame or matrix with first two columns containing locations of a point pattern.

OPTIONAL ARGUMENTS

obj2 an object of class `"spp"` representing a spatial point pattern, or a data frame or matrix with first two columns containing the origins from which the distances to the points in `obj1` are to be computed. See DETAILS for computation of `Fhat`.

nx,ny if `obj2` is missing, a grid of size (`nx` by `ny`) is computed to be used as set of origins. Defaults to the square root of the total number of points in `obj1`.

dist.fhat distances at which `Fhat` values are desired. See DETAILS. By default, all distances between `obj1` and `obj2` will be used.

plot.it logical flag: should the result be plotted? Defaults to TRUE.

VALUE

a matrix with two columns. The first column contains the distances at which `Fhat` was computed. The second column contains the corresponding `Fhat` values.

SIDE EFFECTS

If `plot.it=TRUE`, a plot of `Fhat` versus distance is produced.

DETAILS

`Fhat` provides an estimate of `F(y)`, the proportion of points on a grid (`obj2`) within distance `y` of the nearest point in the original pattern (`obj1`).

$$\hat{F}(y) = m^{-1} \sum_{e_i < y} 1$$

where m is ($nx * ny$) and e_i is the distance between the ith origin and its nearest neighbor in obj1.

For a completely spatially random process without edge effects, the theoretical distribution of $F(y)$ is:
$$F(y) = 1 - exp(-\pi \lambda y^2)$$
where λ is the intensity, the number of points per unit area.

If obj2 is not supplied, an origin grid with dimension (nx x ny) is created on the same area as the original data. The distances between each origin in obj2 and its nearest neighbor in obj1 are computed using find.neighbor.

REFERENCES

Diggle, Peter J. (1983). *Statistical Analysis of Spatial Point Patterns*. Academic Press, London.

SEE ALSO

Ghat, find.neighbor.

EXAMPLES

lans.fhat <- Fhat(lansing)

---

find.neighbor	Find the Nearest Neighbors of a Point	**find.neighbor**

DESCRIPTION

Find the k nearest neighbors of a vector x in a matrix of data contained in an object of class "quad.tree".

USAGE

find.neighbor(x, quadtree=quad.tree(x), k=1, metric="euclidean",
        max.dist=NULL)

REQUIRED ARGUMENTS

x a vector (or matrix) containing the multidimensional point(s) at which the nearest neighbors are desired. The vector must have the same number of elements as the number of columns in the numeric matrix used to construct quadtree. If a matrix is used, the matrix must have the same number of columns as the numeric matrix used to construct quadtree, and nearest neighbors are found for each row in the matrix.

OPTIONAL ARGUMENTS

quadtree an object of class "quad.tree" containing the sorted matrix of data for which a nearest neighbor search is desired. Defaults to quad.tree(x) if x is a matrix but it is required when x is a vector.

k the number of nearest neighbors to be found. If the data x is the same data that was used to construct the "quad.tree" object, then k = 1 results in each element having itself as its own nearest neighbor.

metric a character string giving the metric to be used when finding "nearest" neighbors. Partial matching is allowed. Possible values are: "euclidean", "city block", and "maximum absolute value" for the l_2, l_1, and l_∞ norm, respectively. For two vectors x and y, these are defined as:

$$l_1 = \sum_i |x_i - y_i|,$$

$$l_2 = \sqrt{\sum_i (x_i - y_i)\,2)},$$

$$l_\infty = \max_i |x_i - y_i|$$

max.dist   if max.dist is given, argument k is ignored, and all of the neighbors within distance max.dist of each row in x are found.

VALUE

a matrix with three named columns:

index1   if x is a matrix, the row in x for this nearest neighbor. If x is not a matrix, the value 1.

index2   the row in the matrix from which the quad tree was formed for this nearest neighbor. If the quad tree was formed from a matrix y, then x[index1[i],] and y[index2[i],] are neighbors.

distances   the corresponding nearest neighbor distances.

DETAILS

An efficient recursive algorithm is used to find all nearest neighbors. First the quad tree is traversed to find the leaf with medians nearest the point for which neighbors are desired. Then all observations in the leaf are searched to find nearest neighbors. Finally, if necessary, adjoining leaves are searched for nearest neighbors.

REFERENCES

Friedman, J., Bentley, J. L., and Finkel, R. A. (1977). An algorithm for finding best matches in logarithmic expected time. *ACM Transactions on Mathematical Software* **3**, 209-226.

SEE ALSO

quad.tree.

EXAMPLES

```
x <- cbind(sids$easting, sids$northing)
sids.nhbr <- find.neighbor(x, max.dist = 30)
```

Ghat	EDF of Point-to-Point Nearest Neighbor Distances	**Ghat**

DESCRIPTION

Computes and plots the empirical distribution function (EDF) of the point-to-point nearest neighbor distances for a spatial point pattern.

USAGE

```
Ghat(object, dist.ghat=all.dists, plot.it=T)
```

REQUIRED ARGUMENTS

object   an object of class "spp" representing a spatial point pattern, or a data frame or matrix with first two columns containing locations of a point pattern.

OPTIONAL ARGUMENTS

dist.ghat   a numeric vector containing the distances for which Ghat will be computed. Default is to compute Ghat at every neighbor distance. See DETAILS for definition of Ghat.

plot.it   logical flag: should the resulting EDF be plotted? Defaults to TRUE.

VALUE

a matrix with two columns containing the neighbor distances and corresponding values for the EDF.

SIDE EFFECTS

if plot.it=TRUE, a plot of Ghat versus distance of will be produced.

DETAILS

Ghat provides an estimate of G(y), the proportion of points in a spatial point pattern within a distance

y of their nearest neighbor.

$$\hat{G}(y) = n^{-1} \sum_{d_i < y} 1$$

where n is the number of points in `object` and d_i is the distance between point i and its nearest neighbor. The nearest neighbor distances are computed using `find.neighbor`.

For a completely spatially random process without edge effects the theoretical distribution of G(y) is:

$$G(y) = 1 - exp(-\pi \lambda y^2)$$

where λ is the intensity, the number of points per unit area.

REFERENCES
Diggle, Peter J. (1983). *Statistical Analysis of Spatial Point Patterns.* Academic Press, London.

SEE ALSO
Fhat, find.neighbor.

EXAMPLES
lans.ghat <- Ghat(lansing)

---

hex.legend	Add a Legend Hexagonal Lattice Plot	**hex.legend**

DESCRIPTION
Adds a legend to the plot of an object of class `"hexbin"`. To be called by the corresponding `plot` method. The legend represents the lattice counts. This function is support for `plot.hexbin`, it is not meant to be called directly by the user.

USAGE
```
hex.legend(col, at, labels=format(round(at)), legend.lab="Counts",
    height=3, lcex=1, width=1, inner, style="grayscale", minarea=0.04,
    maxarea=0.8, maxcount, density=-1)
```

REQUIRED ARGUMENTS
col numeric vector of color numbers used to color the different levels in the color legend.

at numeric vector specifying the values at which the colors change.

OPTIONAL ARGUMENTS
labels a character vector of tick labels marking the value of at.

legend.lab a character string denoting a title for the legend itself. The default is "Counts".

height total height of the legend in inches. The default is 3 inches.

lcex numeric constant. Character expansion for the legend text. Default is 1.

width numeric constant giving the desired width of the legend. Default is 1.

inner numeric constant needed to determine the size of the hexagons. Default is computed as a function of width. (Inner diameter in inches.)

style character string. One of 5 possible styles for plotting the hexagons. This must be one of "grayscale","lattice","centroids","nested.lattice", or "nested.centroids".

minarea fraction of cell area for the lowest count. Default is 0.04 or 4%.

maxarea fraction of the cell area for the largest count. Default is 80%.

maxcount cells with more counts than maxcount are ignored.

density same as the corresponding argument in the function polygon. If density is zero, no shading will occur. If density is negative, the hexagons will be filled solidly.

SIDE EFFECTS
A legend is added to the current plot.

DETAILS

This function must not be used as a standalone function. It is meant to be called from `plot.hexbin`.

SEE ALSO

`plot.hexbin`, `hexagons`

| **hexagons** | Add Hexagonal Cells to Plot of `"hexbin"` Object | **hexagons** |

DESCRIPTION

Plots hexagonal cells defined by an object of class `"hexbin"`. This function is primarily support for the function `plot.hexbin`.

USAGE

```
hexagons(bin, style="grayscale", cuts=16, col.regions=
    trellis.par.get("regions")$col, at, mincount=1,
    rmaxcount=max(bin$count), minarea=0.04, maxarea=0.8,
    density=-1, border=F)
```

REQUIRED ARGUMENTS

bin   an object of class `"hexbin"`.

OPTIONAL ARGUMENTS

style   One of the strings `"grayscale"`,`"lattice"`,`"centroids"`,`"nested.lattice"`, or `"nested.centroids"` representing the style of plot desired. See the DETAILS section for the description of each option. Partial string matching is allowed.

cuts   gives the number of levels to divide up the range of the number of counts for each bin. This is used for generating the `at` vector. How many different colors are used will depend, to an extent, on this parameter. Default is `16`.

col.regions   numeric vector integers. These correspond to the colors in the graphics device colormap that are to be used to color the hexagons. Different color devices have more or less adequate colormaps. Using a device defined with a call to `trellis.device` will ensure that a colormap close to adequate is used. See the DETAILS section for ways to select other colormaps.

at   numeric vector giving the breaks at which to divide the range of cell counts. This allows more flexibility in the coloring by allowing non-equally spaced color intervals. If `at` is omitted, it is computed as `seq(min(bin$count),max(bin$count),length=cuts)`.

mincount   cells with counts smaller that `mincount` are not shown.

maxcount   cells with counts larger that `maxcount` are not shown.

minarea   minimum symbol area as a fraction of the binning cell. The fraction of cell area for the lowest count. If small hexagons are hard to see, increase `minarea`.

maxarea   maximum symbol area as a fraction of the binning cell. The fraction of the cell area for the largest count.

density   density of shading lines in lines per inch. If density is zero, no shading will occur. If density is negative, the polygon will be filled solidly using the device-dependent polygon filling algorithm. This is the same as the `density` argument to the S-PLUS function `polygon`.

border   logical flag: should the border of the hexagons be drawn?

SIDE EFFECTS

Adds hexagons to the plot.

DETAILS

The five plotting styles are:

`style="grayscale"` A smoothly varying color mapping of the counts is determined from the values in `cuts`, `at`, and `col.regions`. The best use of this option requires that the plotting device is activated through a call to the S-PLUS function `trellis.device`. This ensures that an adequate color map is the default although other devices as well as customized colormaps can be provided by the user.

`style="lattice"` or `"centroids"` Plots the hexagons in sizes proportional to cell counts. The `"lattice"` option places the hexagons at the lattice centers. In some cases, the regularity of this structure may be visually overwhelming. In those cases, the user should use the `"centroids"` option which places the hexagons at their centers of mass. This results in the breaking of the regularity of the lattice structure thereby placing the focus on other properties of the data. In all cases the hexagons will not plot outside the cell unless `maxarea > 1`.

`style="nested.lattice"` and `"nested.centroids"` Two overlaying hexagons are plotted: a background hexagon with area covering the full hexagon's and color proportional to the cell count in powers of 10 and a foreground hexagon with area proportional to `log10(count)-floor(log10(count))`. When `style="nested.centroids"` counts <10 are plotted and the centers of the plotted hexagons are placed at their centers of mass. The outside color encodes hexagon size within color contours representing powers of 10. Different color schemes give different effects including 3-D illusions

The hexagon sizes are scaled proportionally to cell counts by scaling the counts between `mincount` and `maxcount` and then mapping to areas between `minarea` and `maxarea`.

PERCEPTUAL CONSIDERATIONS

Plotting the symbols near the center of mass is not only more accurate, it helps reduce the visual dominance of the lattice structure. Of course higher resolution binning reduces the possible distance between the center of mass for a bin and the bin center. When symbols nearly fill their bin, the plot appears to vibrate. This can be partially controlled by reducing `maxarea` or by reducing contrast.

The local background influences color interpretation. Having defined color breaks to focus attention on specific contours can help. See `nested` options.

REFERENCES

Carr, D. B. (1991). Looking at large data sets using binned data plots. In *Computing and Graphics in Statistics*. A. Buja and P. Tukey, eds. Springer-Verlag, New York. pp. 7-39.

SEE ALSO

`hexbin`, `plot.hexbin`, `identify.hexbin`, `hex.legend`, `smooth.hexbin`, `erode.hexbin`.

EXAMPLES

```
# A better approach uses plot.hexbin(bin) and controls the plot shape
bin <- hexbin(x,y)
plot(range(bin$x),range(bin$y),type='n')
hexagons(bin)
```

| **hexbin** | Bivariate Binning into Hexagonal Cells | **hexbin** |

DESCRIPTION

Creates an object of class `"hexbin"`. Its basic components are a cell identifier and a count of the points falling into each occupied cell.

USAGE

```
hexbin(x, y, xbins=30, shape=1, xlim=range(x), ylim=range(y))
```

REQUIRED ARGUMENTS

x numeric vector. Usually the first (horizontal) coordinate of bivariate data to be binned.

y numeric vector. Usually the second (vertical) coordinate of bivariate data to be binned.

OPTIONAL ARGUMENTS

xbins number of bins partitioning the range of x values.

shape height to width ratio for the resulting plotting region. This parameter is used in determining the number of bins in the y-direction given `xbins`, the number of bins in the x-direction. The default value `shape = 1` makes the hexagons appear equal-sided when plotted.

xlim the horizontal limits of the binning region in units of x. By default these are the minimum and maximum values of x.

ylim the vertical limits of the binning region in y units. This defaults to the minimum and maximum values of y.

VALUE

an object of class `"hexbin"`. This is a data frame with the following columns:

cell vector of cell identifiers that can be mapped into the bin centers in data units

count a vector with the points count for each corresponding cell.

xcenter the x center of mass (average of x values) for the cell.

ycenter the y center of mass (average of y values) for the cell.

The returned data frame has attributes:

class the class of the returned object, `"hexbin"`.

call the original call to `hexbin` which generated the object.

xbins number of hexagons along the x axis. Same as the input value for `xbins`. Hexagons inner diameter equals `diff(xlim)/xbins` in x units

dims the i-th and j-th limits of `count` if treated as a matrix (`count[i,j]`).

xlim same as the input value for `xlim`.

ylim same as the input value for `ylim`.

shape same as the input value for `shape`.

DETAILS

Returns counts for non-empty cells. The plot shape must be maintained for hexagons to appear with equal sides. Calculations are in single precision.

REFERENCES

Carr, D. B., Littlefield, R. J., Nicholson, W. L. and Littlefield, J. S. (1987). Scatterplot matrix techniques for large N. *Journal American Statistical Association* **83**, 424-436.

SEE ALSO

summary.hexbin, plot.hexbin, identify.hexbin, smooth.hexbin, erode.hexbin, cell2xy, xy2cell.

EXAMPLES

```
x <- rnorm(10000)
y <- rnorm(10000)

bin1 <- hexbin(x,y)
trellis.device(motif)
plot(bin1,style="nested.centroids")

# Lower resolution binning and overplotting with counts
bin2 <- hexbin(x,y,xbins=25)
binpar <- plot(bin2, style="latt" ,minarea=1, maxarea=1, density=0,
      border=T)
oldpar <- par(binpar)      # reset graphics to the plot on the screen
xy <- cell2xy(bin2)
text(xy$x, xy$y, as.character(bin2$count), adj=.5, cex=.3)
par(oldpar)        # restore old graphics parameters

maples <- lansing[lansing$species=="maple",]
plot(maples$x,maples$y)
maple.bin <- hexbin(maples$x,maples$y,xbins=10)
hexagons(maple.bin,border=T,dens=0)
```

---

identify.hexbin   Identify Points On a Hexagonal Binned Plot   **identify.hexbin**

DESCRIPTION

Interactively identifies points on the plot of a `"hexbin"` object if a suitable graphics device is being used.

USAGE

```
identify.hexbin(x, use.pars=par(), labels, tolerance=0.2,
             offset=0, ...)
```

REQUIRED ARGUMENTS

x a `"hexbin"` object. This must have components `count`, `xcenter`, and `ycenter`.

use.pars list of graphical parameters returned when using `plot.hexbin` to plot x before identifying.

OPTIONAL ARGUMENTS

labels vector giving labels for each of the hexagons. If supplied, it must be the same length as `dim(x)[1]`. The default is to use `x$count`, the cell counts.

tolerance points within `tolerance` inches from the pointer will be identified. The default is to identify a point if it is within a tenth of an inch of the pointer.

offset offset for individual labels. The default implies that the text string for identification will be placed right at the cell centers. See this parameter's definition in the help file for `identify.default`.

... optional arguments taken from the argument list for `identify.default`.

VALUE

indices of the identified points.

SIDE EFFECTS

labels are placed on the current plot if `plot=TRUE`, the default for `identify.default`.

DETAILS

This function is a method for the generic function `identify` for class `"hexbin"`. It can be invoked by calling `identify` for an object of the appropriate class, or directly by calling `identify.hexbin` regardless of the class of the object.

SEE ALSO

`identify`, `plot.hexbin`, `hexagons`.

EXAMPLES

```
bin1 <- hexbin(x,y)
bin1pars <- plot(bin1)  # Save the parameters used to plot the hexagons
identify(bin1, bin1pars)
```

---

identify.vgram.cloud   Identify Points On a Variogram Cloud Plot   **identify.vgram.cloud**

DESCRIPTION

Interactively identifies points on a variogram cloud plot if a suitable graphics device is being used.

USAGE

`identify.vgram.cloud(x, labels, plot=T, ...)`

REQUIRED ARGUMENTS

x a `"vgram.cloud"` object. This must have components `distance`, `gamma`, `iindex` and `jindex`.

OPTIONAL ARGUMENTS

labels vector giving labels for each of the points. If supplied, this must have the same length as `x$distance`. The default is the row indices for the two values making up the point separated by a comma.

plot if `TRUE`, `identify` plots the labels of the points identified. In any case, the subscripts are returned.

... optional arguments taken from the argument list for `identify.default`.

VALUE

indices of the identified points.

SIDE EFFECTS

labels are placed on the current plot if `plot = TRUE`, the default for `identify.default`.

DETAILS

This function is a method for the generic function `identify` for class `"vgram.cloud"`. It can be invoked by calling `identify` for an object of the appropriate class, or directly by calling `identify.vgram.cloud` regardless of the class of the object.

SEE ALSO

`identify`, `plot.vgram.cloud`, `variogram.cloud`.

EXAMPLES

```
vc.coal <- variogram.cloud(coal ~ loc(x,y), data=coal.ash)
plot(vc.coal)
identify(vc.coal)
```

`intensity`	Estimate the Intensity of a Spatial Point Pattern	`intensity`

DESCRIPTION

Compute a smooth estimate of intensity given an object of class `"spp"`.

USAGE

```
intensity(object, method = "kernel", bw, boundary = bbox(object), nx,
          ny, edge = F, ...)
```

REQUIRED ARGUMENTS

object a spatial point pattern object. An object of class `"spp"`.

bw bandwidth parameter for methods `"kernel"` and `"gauss2d"`. This should be a numeric value. The same value is used each of the x or the y directions. This value needs to be determined by the user mostly by trial-and-error. A good starting value might be $1/4$ of the diameter of the boundary region.

OPTIONAL ARGUMENTS

method a character string with one of four possible methods for providing an estimate of intensity for object. This must be one of `"basic"`, `"kernel"`, `"binning"`, or `"gauss2d"`. See the DETAILS section for each definition. Defaults to `"kernel"` which applies a quartic kernel to the two-dimensional process. Partial string matching is allowed.

boundary points defining the boundary polygon for the spatial point pattern. This version accepts only rectangles, for which boundary should be given as a list with named components `"x"` and `"y"` denoting the corners of the rectangular region. For example, for the unit square the boundary could be given as bbox(x=c(0,1),y=c(0,1)), the bounding box of two diagonally opposed points. Defaults to a rectangle covering the range of all the points in object.

nx integer. Number of bins in the X direction, or number of X-points in the grid used by the kernel estimators. Defaults to twice the square root of the total number of points in object.

ny integer. Number of bins in the Y direction, or number of Y-points in the grid used by the kernel estimators. Defaults to twice the square root of the total number of points in object.

edge logical flag: should we correct for edge effects? If TRUE, then a toroidal correction is applied. This is equivalent to reflecting the mapped pattern eight times around the one given by object so it may slow down computation. Default is FALSE.

... Other parameters may be given to be passed on to the loess smoother. In particular:

span loess smoothing parameter, used by method =`"binning"` to smooth the bin counts. See the loess help file for more information. Defaults to 0.75 with higher values producing smoother estimates.

VALUE

if method=`"basic"` a number is returned. In all other cases the return value is a list with 4 components as follows:

x,y numeric vectors containing the gridded values in the horizontal and vertical directions respectively.

z estimates of intensity corresponding to x, y respectively.

bw the input bandwidth parameter.

The returned list also has an attribute `"call"` describing the call that generated the resulting object.

DETAILS

The returned value may be given to the S-PLUS function image for a graphical display of intensity over the area enclosed by boundary.

When method=`"binning"`, the points are binned into a nx by ny grid of counts and these are smoothed out using a call to the S-PLUS function loess.

When method=`"kernel"`, a quartic kernel is placed at each cross point on an overlaid nx by ny grid and a two-dimensional kernel smoothing operation performed. This kernel is computed as the product of two univariate quartic kernels, and so it is not rotationally invariant. See the S+SPATIALSTATS function `kern2d` for more information on the formulae used for these computations.

When `method ="gauss2d"`, a gaussian kernel is used instead of the quartic of above.

WARNING

When `edge` is set to `TRUE`, a lot more memory will be required. Increasing `options("object.size")` might be necessary.

REFERENCES

Diggle, Peter J. (1983). *Statistical Analysis of Spatial Point Patterns.* Academic Press, London.

Venables, W. N. and Ripley, B. D. (1994). *Modern Applied Statistics with S-Plus.* Springer-Verlag, New York.

SEE ALSO

`kern2d`, `loess`.

EXAMPLES

```
image(intensity(lansing, "binning", nx=50, ny=50, span=.1))
image(intensity(lansing, bw=.3))
```

Kenv	Compute Simulations of `Khat`	**Kenv**

DESCRIPTION

Computes `Khat` (`Lhat`) for simulations of point processes. Returns upper and lower bounds, as well as the average of all simulated values.

USAGE

```
Kenv(object, nsims=100, maxdist=<<see below>>, ndist=100,
    process="binomial", boundary=bbox(object), add=T, ...)
Lenv(object, nsims=100, maxdist=<<see below>>, ndist=100,
    process="binomial", boundary=bbox(object), add=T, ...)
```

REQUIRED ARGUMENTS

object   a spatial point pattern object. An object of class `"spp"`.

OPTIONAL ARGUMENTS

nsims   integer. Number of desired simulations.

maxdist   numeric value indicating the maximum distance at which `Khat` (or `Lhat`) should be estimated. Defaults to half the length of a diagonal of the sample's bounding box.

ndist   desired number of default distances at which to compute `Khat` (or `Lhat`). Default is `100`.

process   a character string with one of five possible processes for the spatial arrangement of the resulting pattern. This must be one of `"binomial"`, `"poisson"`, `"cluster"`, `"Strauss"`, or `"SSI"`. See the help file for `make.pattern` for information on parameters for each process.

add   logical flag: should the envelope be added to an already existing plot of `Khat` (or `Lhat` for `Lenv`)? Defaults to `TRUE`.

...   other parameters as needed by the requested process.

VALUE

invisibly returns a list with 4 numeric vectors each representing:

dist the distances at which all values were computed.
lower the minimum of all resulting `Khat` (or `Lhat` for `Lenv`) for the simulations.
upper the maximum of all resulting `Khat` (or `Lhat` for `Lenv`) for the simulations.
average the average of all resulting `Khat` (or `Lhat` for `Lenv`) for the simulations.

SIDE EFFECTS

if `add=TRUE` an envelope is added to an existing plot of `Khat`.

SEE ALSO

`Khat`, `Lhat`, `make.pattern`.

EXAMPLES

```
Khat(bramble)
Kenv(bramble,nsims=50)
Lhat(lansing.spp)
Lenv(lansing.spp)
```

---

kern2d	Kernel Smoothing of a Two-Dimensional Density	**kern2d**

DESCRIPTION

Smooths a two-dimensional process using kernels.

USAGE

```
kern2d(x, y, bw=<<see below>>, nx=25, ny=25,
       lims=c(range(x), range(y)), kernfun=dnorm)
```

REQUIRED ARGUMENTS

x numeric vector with the X-coordinates of the two-dimensional process
y numeric vector with the Y-coordinates of the two-dimensional process. This must be the same length as x.

OPTIONAL ARGUMENTS

bw bandwidth parameter. This should be a numeric vector with 2 components, giving the values desired for the bandwidth in each of the x or the y directions. If only one number, then the same value will be repeated for both directions. See the DETAILS section for its default.
nx integer. Number of X-points in the grid used by the kernel estimators. Defaults to 25.
ny integer. Number of Y-points in the grid used by the kernel estimators. Defaults to 25.
lims numeric vector with 4 components defining the bounding box of the two-dimensional process to smooth.
kernfun the kernel function. Use the function `dnorm` for the Gaussian kernel, or `kernquart` for the quartic. This can be any function that takes a one-dimensional vector and returns its smoothed version.

VALUE

a list with 4 components as follows:
x,y numeric vectors containing the gridded values in the horizontal and vertical directions respectively.
z estimates of intensity corresponding to x, y respectively.
bw the input or estimated bandwidth parameter.

DETAILS

The returned value may be given to the function `image` for a graphical display of intensity over the area enclosed by `boundary`.

The bandwidth parameter default value is determined using Venables and Ripley's (1994) suggested "rule of thumb" for the Inter Quantile Range R and the Gaussian kernel of bandwidth the standard deviation,

```
bw <- 4*1.06*min(sqrt(var(x)), R/1.34)*length(x)^(-1/5)
```

Ideally, the user should try several values before deciding on an useful bandwidth, perhaps starting with the default above.

This function is based on Venables and Ripley's `kde2d` function.

REFERENCES

Venables, W. N. and Ripley, B. D. (1994). *Modern Applied Statistics with S-Plus.* Springer-Verlag, New York.

SEE ALSO

`intensity, dnorm.`

EXAMPLES

```
plot(geyser$duration, geyser$waiting, type="n")
image(kern2d(geyser$duration, geyser$waiting))
points(geyser$duration, geyser$waiting)
```

Khat	Ripley's K Function for a Spatial Point Pattern Object	**Khat**

DESCRIPTION

Calculates `K(t)`, Ripley's K function for a spatial point pattern.

USAGE

```
Khat(object, maxdist=<<see below>>, ndist=100, boundary=bbox(object),
     plot.it=T)
```

REQUIRED ARGUMENTS

object   a spatial point pattern object. An object of class `"spp"`.

OPTIONAL ARGUMENTS

maxdist   numeric value indicating the maximum distance at which `Khat` should be estimated. Defaults to half the length of a diagonal of the sample's bounding box.

ndist   desired number of default distances at which to compute `Khat`. Default is `100`. The distances for which `Khat` will be estimated are calculated as `seq(0,maxdist,ndist)`, both `maxdist` and `ndist` will change if not reasonable for the given `object`.

boundary   points defining the boundary polygon for the spatial point pattern. This version accepts only rectangles, for which `boundary` should be given as a list with named components `"x"` and `"y"` denoting the corners of the rectangular region. For example, for the unit square the boundary could be given as `bbox(x=c(0,1),y=c(0,1))`, the bounding box of two diagonally opposed points. Defaults to a rectangle covering the range of points.

plot.it   logical flag: should the resulting K-estimates be plotted? Default is `TRUE`.

VALUE

a list containing components :

values   a two column matrix. The first column, named `dist`, contains the distances at which `Khat` was computed, and the second column, named `Khat`, contains the values of `K(dist)`.

ndist number of distances returned. This could be smaller than its input value if the extent of the distances is too large.

mindist minimum distance between any pair of points.

maxdev maximum deviation from K(t)=t. See DETAILS.

SIDE EFFECTS

if plot.it=TRUE, a plot of the value of K(t) against distance will be produced on the current graphics device.

DETAILS

Khat computes Ripley's (1976) estimate of K(t) for a spatial point pattern:

$$K(t) = \lambda^{-1} E[\textit{number of events} \leq \textit{distance t of an arbitrary event}].$$

where λ is the intensity of the spatial point pattern.

The theoretical K-function for a Poisson (completely spatially random) process is $K(t) = \pi t^2$, so $L(t) = \sqrt{K(t)/\pi}$ is equal to t, the distances. The default plots K(t) versus t. See function Lhat for estimation of L(t).

REFERENCES

Ripley, Brian D. (1976). The second-order analysis of stationary point processes. *Journal of Applied Probability,* **13**,255-266.

SEE ALSO

Kenv, Lhat.

EXAMPLES

```
lansing.spp <- as.spp(lansing)
lansing.khat <- Khat(lansing.spp)
Khat(wheat)
abline(0,1)
```

---

krige	Ordinary and Universal Kriging	**krige**

DESCRIPTION

Performs ordinary or universal kriging for two dimensional spatial data. The function predict.krige can then be called to compute interpolation surfaces and prediction errors.

USAGE

```
krige(formula, data=sys.parent(), subset, na.action=na.fail,
      covfun, nc=1000, ...)
```

REQUIRED ARGUMENTS

formula a formula describing the kriging variable and the spatial location variables and optionally a polynomial trend surface. Its simplest form is:

```
z ~ loc(x,y)
```

where z is the kriging variable and x and y are the spatial locations i.e. z[i] is observed at the location (x[i],y[i]). The right hand side must contain a call to the function loc. A polynomial trend surface is of the form:

```
z ~ loc(x,y) + x + y + x^2 + y^2
```

The polynomial must be in the same variables as the first two arguments used in the `loc` function. A constant term is always fit. All terms on the right hand side must be entered with a + sign. The `loc` call can include arguments `angle` and `ratio` to correct for geometric anisotropy, see the `loc` help file. Note that an evaluated `loc` object cannot be used in `formula`.

`covfun` a function that returns the distanced based covariance between two points. The first argument to the function must be the distance. Additional parameters will be passed through the . . . .

OPTIONAL ARGUMENTS

`data` an optional data frame in which to find the objects mentioned in `formula`.

`subset` expression saying which subset of the rows of the data should be used in the fit. This can be a logical vector (which is replicated to have length equal to the number of observations), or a numeric vector indicating which observation numbers are to be included, or a character vector of the row names to be included.

`na.action` a function to filter missing data. This is applied to the data in `formula` after any `subset` argument has been used. The default (with `na.fail`) is to create an error if any missing values are found. A possible alternative is `na.omit`, which deletes observations that contain one or more missing values.

`nc` the number of points to use internally in approximating the distance based covariance.

`...` additional named arguments can be passed to `covfun`.

VALUE

an object of class `"krige"` with components:

`x` the first spatial location vector i.e. the first argument in `loc` function call in `formula`.

`y` the second spatial location vector i.e. the second argument in `loc` function call in `formula`.

`coefficients` the vector of coefficients for the trend surface. These are for the polynomial based on the scaled spatial location vectors (see the DETAILS section).

`residuals` the vector of residuals from the trend surface.

`call` an image of the call that produced the object.

Other components are included that will be used by `predict.krige` for computing interpolations.

DETAILS

The kriging system is solved using generalized least squares (see Ripley, 1981). The polynomial terms are scaled to (-1, 1) internally to avoid numeric problems, the `coefficients` component returned is for these scaled terms.

This implementation of kriging does not handle multiple observations at a point.

Methods for objects of class `"krige"` include `predict` and `print`.

REFERENCES

Cressie, Noel A. C. (1993). *Statistics for Spatial Data,* Revised Edition. Wiley, New York.

Ripley, Brian D. (1981). *Spatial Statistics.* Wiley, New York

SEE ALSO

`exp.cov`, `loc`, `predict.krige`.

EXAMPLES

```
# krige the Coal Ash data with a quadratic trend in the x direction
#   using a spherical covariance function:
kcoal <- krige(coal ~ loc(x, y) + x + x^2, data = coal.ash,
        covfun = spher.cov, range = 4.31, sill = 0.14, nugget = 0.89)
# predictions over default 30 x 30 grid
```

```
pcoal <- predict(kcoal)
# plot prediction surface
wireframe(fit ~ x * y, data = pcoal,
        screen = list(z = 300, x = -60, y = 0), drape = T)
```

Lhat	Ripley's K Function for a Spatial Point Pattern Object	**Lhat**

DESCRIPTION

Calculates `L(t)=sqrt(K(t)/pi)`, where `K(t)` is Ripley's K function for a spatial point pattern and `L(t)` is linear for a completely random point process.

USAGE

```
Lhat(object, maxdist=<<see below>>, ndist=100, boundary=bbox(object),
    plot.it=T)
```

REQUIRED ARGUMENTS

object a spatial point pattern object. An object of class `"spp"`.

OPTIONAL ARGUMENTS

maxdist numeric value indicating the maximum distance at which `Lhat` should be estimated. Defaults to half the length of a diagonal of the sample's bounding box.

ndist desired number of default distances at which to compute `Lhat`. Default is `100`. The distances for which `Lhat` will be estimated are calculated as `seq(0,maxdist,ndist)`, both `maxdist` and `ndist` will change if not reasonable for the given `object`.

boundary points defining the boundary polygon for the spatial point pattern. This version accepts only rectangles, for which `boundary` should be given as a list with named components `"x"` and `"y"` denoting the corners of the rectangular region. For example, for the unit square the boundary could be given as `bbox(x=c(0,1),y=c(0,1))`, the bounding box of two diagonally opposed points. Defaults to a rectangle covering the range of points.

plot.it logical flag: should the resulting `K`-estimates be plotted? Default is `TRUE`.

VALUE

a list containing components :

values a two column matrix. The first column, called `dist`, contains the distances at which `Lhat` was computed, and the second column, called `Lhat`, contains the values of `L(dist)`.

ndist number of distances returned. This could be smaller than its input value if the extent of the distances is too large.

mindist minimum distance between any pair of points.

maxdev maximum deviation from `L(t)=t`. See DETAILS.

SIDE EFFECTS

if `plot.it=TRUE`, a plot of the value of `L(t)` against distance will be produced on the current graphics device.

DETAILS

`Khat` computes Ripley's (1976) estimate of K(t) for a spatial point pattern:

$$K(t) = \lambda^{-1} E[\textit{number of events} \leq \textit{distance t of an arbitrary event}].$$

where λ is the intensity of the spatial point pattern.

The theoretical K-function for a Poisson (completely spatially random) process is $K(t) = \pi t^2$, so $L(t) = \sqrt{K(t)/\pi}$ is equal to t, the distances. The default plots `L(t)` versus t which should approximate a straight line for a homogeneous process with no spatial dependence. See function `Khat` for estimation of `K(t)`.

REFERENCES
Ripley, Brian D. (1976). The second-order analysis of stationary point processes. *Journal of Applied Probability* **13**,255-266.

SEE ALSO
Lenv, Khat.

EXAMPLES
```
lansing.spp <- as.spp(lansing)
lansing.khat <- Lhat(lansing.spp)

Lhat(wheat)
abline(0,1)
```

---

loc	Correct Spatial Locations for Geometric Anisotropy	**loc**

DESCRIPTION
The input spatial locations are rotated, scaled and rotated back to correct for geometric anisotropy defined by angle of anisotropy and ratio. This function is usually used in the formula for estimating a variogram.

USAGE
```
loc(x, y, angle=0, ratio=1)
```

REQUIRED ARGUMENTS
x   numeric vector of x locations, must be the same length as y. Alternatively x can be a two column matrix containing both the x and y locations.

OPTIONAL ARGUMENTS
y   numeric vector of y locations, must be the same length as x.

angle   an angle measured clockwise from North in degrees.

ratio   a numeric value specifying the ratio of anisotropy at angle to the anisotropy orthogonal to angle.

VALUE
a two column matrix of class `"loc"` containing the corrected spatial locations.

DETAILS
The locations are corrected for geometric anisotropy by multipling each location pair ($x[i],y[i]$) by the symmetric 2 x 2 matrix A where `A[1,1]=cos(angle)^2+ratio*sin(angle)^2`, `A[1,2]=(1-ratio)*sin(angle)*cos(angle)` and `A[2,2]=sin(angle)^2+ratio*cos(angle)^2`. See Journel and Huijbregts (1978, pp 179-181).

REFERENCES
Journel, A. G. and Huijbregts, Ch. J. (1978). *Mining Geostatistics.* Academic Press, New York.

SEE ALSO
anisotropy.plot, variogram.

EXAMPLES
```
variogram(ore ~ loc(easting, northing, angle=0, ratio=2), data=iron.ore)
```

`lrt`	Computes a Likelihood Ratio Test	`lrt`

DESCRIPTION

Computes a likelihood ratio test in a maximum likelihood problem.

This function is generic (see `Methods`); method functions can be written to handle specific classes of data. Classes which already have methods for this function include: `slm`

USAGE

```
lrt(object, ...)
```

REQUIRED ARGUMENTS

`object` any object, usually a fitted model object of some kind.

OPTIONAL ARGUMENTS

`...` some methods have additional arguments.

VALUE

the return value can vary with the generic routine, but in general a list containing the computed (asymptotic) chi-squared statistic, its p-value, and the degrees of freedom is returned.

DETAILS

A likelihood ratio test compares two or more likelihoods to examine whether the null hypothesis or an alternative hypothesis provides a significantly better fit to the observed data. Discussion of such tests can be found in most elementary statistics texts.

SEE ALSO

`Methods`, `lrt.slm`.

EXAMPLES

```
object <- slm(sid.ft ~ nwbirths.ft, cov.family=CAR, data=sids, subset= -4,
              spatial.arglist=list(weights=1/sids$births,
          neighbor=sids.neighbor))
lrt(object, 0.12, c(1.3333, NA))
```

`lrt.slm`	Likelihood Ratio Test for Spatial Linear Models	`lrt.slm`

DESCRIPTION

Computes a likelihood ratio test for spatial linear models.

USAGE

```
lrt.slm(object, parameters=NULL, coefficients=NULL, ...)
```

REQUIRED ARGUMENTS

`object` an object of class `"slm"`. The likelihood ratio test is performed by restricting the parameters used to define the model to a smaller class. Using the arguments below, the covariance matrix parameters (but not the scale parameter) or the linear model parameters can be restricted by setting them to known or hypothesized values.

OPTIONAL ARGUMENTS

parameters the fixed values of the covariance matrix parameters specified by the null hypothesis. Vector `parameters` can either be a named vector where the names of the vector are names from `names(object$parameters)` and the values are the values fixed by the null hypothesis. Alternatively, vector `parameters` can be the same length as `object$parameters`, with parameters which are free to vary specified using `NA`'s. If `NULL`, then no covariance matrix parameter is fixed by the null hypothesis and all are free to vary.

coefficients the fixed values of the linear model coefficients specified by the null hypothesis. Vector `coefficients` can either be a named vector where the names of the vector are names from `names(coef(object))` and the values are the values fixed by the null hypothesis. Alternatively, vector `coefficients` can be the same length as `object$coefficients`, with coefficients which are free to vary specified using `NA`'s. If `NULL`, then all coefficients are free to vary and none are fixed by the null hypothesis.

... arguments to be passed on to the routine `slm.nlminb` (if optimization of the log-likelihood under the null hypothesis is required). See routine `slm.nlminb`.

VALUE

a list containing the following elements:

chisquared (asymptotic) chi-squared likelihood ratio statistic.

p.value the (two-sided) p-value for the statistic `chisquared`.

df the degrees of freedom used to compute `p.value`.

parameters the parameters fit under the null hypothesis.

coefficients the coefficients fit under the null hypothesis.

DETAILS

`lrt.slm` is used to perform likelihood ratio tests on the covariance matrix `parameters` and the linear model `coefficients` of a spatial regression model. Elements of either of these vectors can be specified to fixed values, or can be left free to vary. Each (linearly independent) parameter which is restricted to a fixed values contributes one degree of freedom to the chi-squared statistic used in performing the likelihood ratio test.

Likelihood ratio tests compare the optimal likelihood under two hypothesis. In `lrt.slm`, the alternative hypothesis is that all model parameters are free to vary, so the maximum likelihood estimates obtained from `slm` and used as input to `lrt.slm` maximize the likelihood hypothesis under the alternative hypothesis. The null hypothesis is obtained by restricting model parameters to fixed values. If the null hypothesis is true, then twice the log of the ratio of the optimal likelihoods under the null and alternative hypotheses is asymptotically distributed as a chi-squared statistic with degrees of freedom equal to the number of fixed parameters (see Haining, 1990, page 143). Using this chi-squared distribution, the significance level is easily obtained.

Closed form expressions for the optimal linear model coefficients are available if the covariance matrix parameters (except the scale parameter) are known. In this case the likelihood ratio test will be relatively inexpensive to compute. However, if one or more covariance matrix parameters is not known, an iterative algorithm must be used to compute the optimal log-likelihood under the alternative hypothesis. In this case, the optimization algorithm used is identical to the algorithm used by the routine `slm`, and can be quite cpu intensive.

`lrt.slm` does not allow likelihood ratio tests on the scale parameter.

REFERENCES

Haining, R. (1990). *Spatial Data Analysis in the Social and Environmental Sciences.* Cambridge University Press, London.

SEE ALSO

`slm, slm.nlminb`.

EXAMPLES

```
sids.slm <- slm(sid.ft ~ nwbirths.ft, cov.family=CAR, data=sids,
    subset=-4, spatial.arglist=list(weights=1/sids$births,
    neighbor=sids.neighbor))
lrt(sids.slm, 0.12, c(NA, 0))
lrt(sids.slm, 0.12, c("nwbirths.ft"=0))  # another way to specify the test
```

MA	Moving Average Spatial Regression Object	**MA**

DESCRIPTION

An object of class `"cov.family"` containing functions for fitting conditional autoregression models when used as input to the spatial linear models function, `slm`. See `cov.family.object` for a discussion of the attributes contained in the MA object. The discussion here centers around the covariance matrix model which the MA object supports.

DETAILS

Let N denote a neighbor matrix obtained from an object of class `"spatial.neighbor"`, and let Σ denote the covariance matrix of a vector y of spatially correlated dependent variables. Finally, let W denote a diagonal matrix of weights. Then a moving average spatial regression model assumes that

$$\Sigma = (I + \rho N)W(I + \rho N)^T \sigma^2$$

where ρ is a scalar parameter to be estimated, and σ is a scale parameter which is also to be estimated. This model for the covariance matrix can be generalized to multiple matrices N_i using multiple parameters ρ_i as follows:

$$\Sigma = \left[\sum_i (I + \rho_i N_i)\right] W \left[\sum_i (I + \rho_i N_i)\right]^T \sigma^2,$$

where the N_i are specified through component `matrix` of the `"spatial.neighbor"` object (see routine `spatial.neighbor`).

The "regression" aspect of a spatial regression fits the multivariate normal mean vector

$$\mu = E(y|x) = x\beta$$

for unknown parameters β. The multivariate normal likelihood is expressed in terms of the unknowns ρ, σ, and β. The MA object assumes that a profile likelihood for ρ is fit.

The MA model can be expressed as a moving average model for the spatial parameters as follows:

$$y = X\beta + \rho N W^{1/2}\varepsilon + W^{1/2}\varepsilon.$$

This allows one to decompose the sum of squares in y into three components (see Haining, 1990, page 258): 1) the trend, $X\beta$; 2) the noise,

$$W^{1/2}\varepsilon = (I + \rho N)(y - X\beta);$$

and 3) the signal,

$$y - X\beta - W^{1/2}\varepsilon.$$

Function `residual.fun` of the `MA` object computes ε, the standardized residuals, and routine `slm` returns these in component `residuals`. The estimated trend, $X\beta$, is returned by routine `slm` as the fitted values.

Two functions are required to compute the profile likelihood: 1) a function for computing the determinant $|\Sigma|$, and 2) a function for computing the vector $\Sigma^{-1}z$ for arbitrary vector z. When the single neighbor matrix N is symmetric, the determinant can be expressed and efficiently computed as a function of the eigenvalues of N. If N is not symmetric, or if the dimension of Σ is to large (over 150), then sparse matrix routine by Kundert (1988) are use to compute the determinant of Σ. Unlike the SAR, CAR, and other spatial regression models, for moving average models the covariance matrix is not parameterized in terms of its inverse. The matrix N will usually be sparse, however, so $\Sigma^{-1}z$ can be efficiently computed using the Kundert (1988) algorithms. See routine `spatial.solve` for details.

REFERENCES

Haining, R. (1990). *Spatial Data Analysis in the Social and Environmental Sciences.* Cambridge University Press. Cambridge.

Kundert, Kenneth S. and Sangiovanni-Vincentelli, Alberto (1988). A Sparse Linear Equation Solver. Department of EE and CS, University of California, Berkeley.

SEE ALSO

`cov.family.object`, `slm`, `slm.fit`, `slm.nlminb`, `spatial.neighbor`, CAR, SAR, `spatial.multiply`.

---

make.pattern	Generate a Spatial Point Process	**make.pattern**

DESCRIPTION

Generates points in two-dimensional space given their desired spatial distribution.

USAGE

```
make.pattern(n, process="binomial", object, boundary=bbox(x=c(0,1),
        y=c(0,1)), lambda, radius, cpar)
```

REQUIRED ARGUMENTS

n integer denoting the desired number of points in the resulting object.

OPTIONAL ARGUMENTS

process a character string with one of five possible processes for the spatial arrangement of the resulting pattern. This must be one of `"binomial"`, `"poisson"`, `"cluster"`, `"Strauss"`, or `"SSI"`. See the DETAILS section for each definition. Defaults to `"binomial"` for a completely spatially random process conditioned to n points within `boundary`. Partial matching is allowed.

object a spatial point pattern object. An object of class `"spp"`. When this is given, the resulting pattern will have the same n and its `boundary` will be the bounding box of `object`.

boundary points defining the boundary polygon for the spatial point pattern. This version accepts only rectangles, for which `boundary` should be given as a list with named components `"x"` and `"y"` denoting the corners of the rectangular region. For example, for the unit square the boundary could be given as `bbox(x=c(0,1),y=c(0,1))`, the bounding box of two diagonally opposed points. Defaults to `bbox(object)` if `object` is given or to the unit square otherwise.

lambda numeric value representing the desired intensity when `process="poisson"`. n, if given, will be ignored for this option and estimated by the routine.

radius the inhibition distance. This is needed for process `"Strauss"`, `"SSI"` and `"cluster"`. Options `"Strauss"` and `"SSI"` will NOT generate points closer than `radius`. For this reason, this parameter needs to be reasonably small. The exception is when `process="cluster"` in which case it should contain the desired size of the clusters. See DETAILS section for more information.

cpar the inhibition parameter needed when `process="Strauss"`. This parameter is also required if `process="cluster"`. In that case, it represents the intensity of the "parent" Poisson process which will determine the random placement of clusters and their number. See the DETAILS section for more information.

VALUE

an object of class `"spp"` whose n points are distributed according to `process`. If `process="poisson"` results on a process with zero points, the returned value will be a classless matrix with zero rows and a warning will be issued.

DETAILS

The `"binomial"` process option generates a spatially random pattern of n points within the given `boundary`. This is in essence a homogeneous Poisson process conditional on the given number of points n.

The `"poisson"` process option generates a homogeneous Poisson process with intensity `lambda`. This parameter is required for this option.

The `"SSI"` process generates a random pattern where no two points are within the inhibition distance determined by its parameter `radius`. This process is equivalent to sequentially laying down discs of radius `radius` which will not overlap.

The `"Strauss"` process accepts each randomly generated point with probability `cpar^s` where s is the number of existing points within radius `radius` of the potential new point. The parameter `cpar` must be in [0,1] for this process, where `cpar = 0` corresponds to complete inhibition at distances up to `radius`.

The user should exercise caution when determining the value of `radius` for if it is too big in relation to the area defined by `boundary`, the algorithm will run out of possible area to place the subsequent disc and the generation of the desired process may be impossible or very slow.

The option `"cluster"` generates a Poisson cluster process. This is defined by generating a "parent" Poisson process with intensity `cpar` and a "daughter" process of clusters with radii determined by the value of `radius`.

WARNING

If `radius` is too large, it may be impossible or nearly impossible to generate the number of requested points. The call may "hang" in some extreme cases.

REFERENCES

Ripley, Brian D. (1981). *Spatial Statistics*. John Wiley & Sons, New York.

Ripley, Brian D. (1976). The second-order analysis of stationary point processes. *Journal of Applied Probability* **13**,255-266.

SEE ALSO

runif, rnorm, rpois, rbinom.

EXAMPLES

```
# A completely random process in the unit square
rand <- make.pattern(100)
plot(make.pattern(100, process="Strauss", rad = 0.1, c = 0.5))
plot(make.pattern(500, "cluster", rad=20, c=10, boundary=c(0,200,0,200)))
```

| `model.variogram` | Display a Variogram Object and Theoretical Model | `model.variogram` |

DESCRIPTION

Plots an empirical variogram object and displays the fit of a theoretical variogram model on that plot. Optionally allows interactive parameter updates to the theoretical model and displays the new fit.

USAGE

```
model.variogram(object, fun, ..., ask=T, objective.fun=<<see below>>)
```

REQUIRED ARGUMENTS

object   an object the inherits from class `"variogram"` (this include classes `"covariogram"` and `"correlogram"`). The `azimuth` column should have only one level.

fun   a theoretical variogram function (or covariogram or correlogram function, depending on the class of `"object"`). The first argument should be distance. The remaining arguments are considered parameters that can be changed to update the fit of fun to `object`.

OPTIONAL ARGUMENTS

...   the additional arguments to `fun` that do not have default values must be specified here by full name.

ask   a logical value, if `TRUE`, a command line menu is displayed allowing the user to change the values of the parameters to `fun`. After changing a value the plot is updated. If `FALSE`, the data in `object` is plotted, the value of `fun` evaluated at `object$distance`, added to the graph and the function returns.

objective.fun   a function with three arguments, `y`, `yf`, and `n` that gives a measure of the fit of `yf` to `y` with weights `n`. It is used as a measure of fit of `fun` to the data in `object`. The default is the sum of squared residuals, `sum((y-yf)^2)`.

VALUE

invisibly returns a named list of the final parameters used.

DETAILS

This function can be used to fit a variogram or covariogram model "by eye". The value of `objective.fun` is displayed on the plot.

A weighted least squares objective function for variograms (Cressie, 1993, p. 97) is:

```
objective.fun <- function(y,yh,n) sum(n*(y/yh-1)^2)
```

REFERENCES

Cressie, Noel. (1993). *Statistics For Spatial Data,* Revised Edition. Wiley, New York.

SEE ALSO

`correlogram`, `plot.variogram`, `variogram`.

EXAMPLES

```
vg.iron <- variogram(residuals ~ loc(easting, northing), data=iron.ore)
model.variogram(vg.iron, spher.vgram, range=8.7, sill=3.5, nugget=4.8)
```

| **neighbor.grid** | Creates a `"spatial.neighbor"` Object for a Grid | **neighbor.grid** |

DESCRIPTION

Creates an object of class `"spatial.neighbor"` for spatial data arranged in a rectangular lattice, or grid, allowing several definitions of a neighbor pair.

USAGE

```
neighbor.grid(nrow, ncol, neighbor.type="first.order",
              weight.fun=NULL, matrix.fun=NULL, max.horiz.dist=1,
              max.vert.dist=1, use.pattern=F)
```

REQUIRED ARGUMENTS

nrow integer denoting the number or rows in the spatial grid.

ncol integer denoting the number of columns in the spatial grid.

OPTIONAL ARGUMENTS

neighbor.type a character string describing the type of relationship used to define the cells to be a spatial neighbor pair. Partial matching is allowed. Valid values are:

`"none"` - if a null neighbor matrix is to be generated (no neighbors).

`"first.order"` - a neighbor matrix with pattern:

$$\begin{array}{ccc} 0 & 1 & 0 \\ 1 & X & 1 \\ 0 & 1 & 0 \end{array}$$

for interior point X, where regions in the lattice denoted by "1" are neighbors of the region "X" and all other regions are not considered neighbors.

`"second.order"` - a neighbor matrix with pattern:

$$\begin{array}{ccc} 1 & 1 & 1 \\ 1 & X & 1 \\ 1 & 1 & 1 \end{array}$$

for interior point X, where regions in the lattice denoted by "1" are neighbors of the region "X", and all other regions are not considered neighbors.

`"diagonal"` - a neighbor matrix with pattern:

$$\begin{array}{ccc} 1 & 0 & 1 \\ 0 & X & 0 \\ 1 & 0 & 1 \end{array}$$

for interior point X, where regions in the lattice denoted by "1" are neighbors of the region "X", and all other regions are not considered neighbors.

`"hexagonal.in"` - a neighbor matrix with pattern:

$$\begin{array}{ccc} 0 & 1 & 1 \\ 1 & X & 1 \\ 0 & 1 & 1 \end{array}$$

for interior point X, where regions in the lattice denoted by "1" are neighbors of the region "X", and all other regions are not considered neighbors. This pattern may be used with hexagonal grids in which the top row is indented.

`"hexagonal.out"` - a neighbor matrix with pattern:

$$\begin{array}{ccc} 1 & 1 & 0 \\ 1 & X & 1 \\ 1 & 1 & 0 \end{array}$$

for interior point X, where regions in the lattice denoted by "1" are neighbors of the region "X", and all other regions are not considered neighbors. This pattern may be used with hexagonal grids in which the top row is not indented.

`"user"` - the user specifies the neighbors (or the neighbor pattern, see argument use.pattern) using the function arguments weight.fun and matrix.fun.

`weight.fun` a function with calling sequence:

```
weight.fun(row1, col1, row2, col2)
```

which accepts the row and column numbers for two regions (`row1`, `col1`), and (`row2`, `col2`) in the grid and computes a neighbor weight for the two cells. If the weight is zero, the regions are not "neighbors". To reduce the computations, the `max.horiz.dist` and `max.vert.dist` arguments can be used to reduce the number of regions examined to determine if they are neighbors.

`matrix.fun` a function with calling sequence

```
matrix.fun(row1, col1, row2, col2)
```

used to compute the type of the neighbor relationship for the two regions (`row1`, `col1`) and (`row2`, `col2`). Consider a region in which north-south neighbors qualitatively different from east-west neighbors, because, say, of a north-south mountain range. In this and other similar cases, it might make sense to specify different parameterizations for different neighbor relationships. If a small number of different kinds of neighbor relationships can be defined, the `matrix.fun` routine can be used to specify these types. When a positive neighbor weight is found, function `matrix.fun` is called. Its return value specifies the type of the neighbor relationship.

`max.horiz.dist` the maximum horizontal distance on either side of the current grid region to look for neighbors. This argument is used in conjunction with `matrix.fun` or `weight.fun` when searching for neighbors. When defining the cells to search, `max.horiz.dist=1` indicates to search the adjoining "columns" of the grid (as well as the current column), `max.horiz.dist=2` says to search the adjoining two columns, etc.

`max.vert.dist` the maximum vertical distance on either side of the current grid region to look for neighbors. This argument is used in conjunction with `matrix.fun` or `weight.fun` when searching for neighbors. When defining the regions to search, `max.vert.dist=1` indicates to search the adjoining rows of the grid (as well as the current row), `max.vert.dist=2` says to search the adjoining two rows, etc.

`use.pattern` a logical value, if TRUE, `weight.fun` and `matrix.fun`, if present, are called only for a rectangular subset of the grid, where the subset is the minimum subset needed as specified by arguments `max.vert.dist` and `max.horiz.dist`. This small subset is used to set a pattern of weights (and neighbor types), and this pattern is used to generate the `"spatial.neighbor"` object. Notice that when `use.pattern` is TRUE, `weight.fun` and `matrix.fun` are only called for the cells in a `max.vert.dist` x `max.horiz.dist` subset of the grid, saving considerable computational time.

VALUE

an object of class `"spatial.neighbor"`. In the `"spatial.neighbor"` object, regions are labeled consecutively, with the regions down the first column of the grid coming first, the elements down the second column coming next, and so on. For example, in a 5 by 3 grid the region labels are:

$$
\begin{array}{ccc}
1 & 6 & 11 \\
2 & 7 & 12 \\
3 & 8 & 13 \\
4 & 9 & 14 \\
5 & 10 & 15
\end{array}
$$

DETAILS

Objects of class `"spatial.neighbor"` contain spatial weights defining the strength of associations between observations. These objects are required by routines `slm`, `spatial.cor`, and other spatial modeling functions. Methods for efficiently defining these objects on regular and irregular grids are essential.

Function `neighbor.grid` can be used to generate `"spatial.neighbor"` objects on regular grids. Consider a 30 by 30 spatial grid. This grid contains 900 regions and 900 x 899 = 809,100 possible spatial neighbors. For any one region, the number of spatial neighbors is likely to be quite small, usually less than 10, but there are 899 possible spatial neighbors for this region. The arguments `max.vert.dist` and `max.horiz.dist` can be used to limit the search for spatial neighbors to regions "closest" to the region being considered. Further savings in cpu time are possible. It is possible, by setting `use.pattern=TRUE`, to generate an entire neighbor object specifying neighbors for all regions on a grid using a "pattern" of neighbor weights provided by a representative region of the grid, where the pattern weights can be defined by the S-PLUS function `weight.fun`. Once an initial pattern has been defined, this pattern of weights is replicated over the entire region in an obvious way, significantly reducing the computational time. On a 30 by 30 grid and using the default values for `max.vert.dist` and `max.horiz.dist`, only 8*900 calls to `weight.fun` need be made (as compared to 899*900 calls otherwise). If `use.pattern=TRUE`, only eight calls (3 * 3 - 1) are made.

SEE ALSO

> `spatial.neighbor`, `spatial.weights`.

EXAMPLES

```
my.grid <- neighbor.grid(nrow=10, ncol=20, neighbor.type="first")

the.weights  <- function(row1, col1, row2, col2)
{
        if(abs(row1 - row2))
                return(1/(abs(col2 - col1) + 1))
        else return(0)
}

two.types <- function(row1, col1, row2, col2) {
        if (col1 < col2) return(1)
        else return (2)
}
zz <- neighbor.grid(10, 20, neighbor.type="user", weight.fun=the.weights,
        matrix.fun=two.types, use.pattern=T)
```

`plot.hexbin`	Plot A Hexagonal Lattice	`plot.hexbin`

DESCRIPTION

> Plots an object of class `"hexbin"` with legend.

USAGE

```
plot.hexbin(bin, style="grayscale", minarea=0.04, maxarea=0.8,
    mincount=1, maxcount=max(bin$count), cuts=min(16, maxcount),
    col.regions=trellis.par.get("regions")$col,
    at=pretty(bin$count, cuts), border=F, density=-1, legend=T,
    legend.width=1, legend.lab="Counts", legend.cex=1, xlab="",
    ylab="", ...)
```

REQUIRED ARGUMENTS

> `bin` an object of class `"hexbin"`.

OPTIONAL ARGUMENTS

> `style` a character string with one of 5 possible styles for plotting the hexagons. This must be one of `"grayscale"`,`"lattice"`,`"centroids"`,`"nested.lattice"`, or `"nested.centroids"`. See the DETAILS section for their descriptions and relative merits. Default is `"grayscale"`. Partial string matching is allowed.

minarea fraction of cell area for the lowest count. Default is 0.04 or 4%.

maxarea fraction of the cell area for the largest count. Default is 80%.

mincount cells with fewer counts than `mincount` are ignored.

maxcount cells with more counts than `maxcount` are ignored.

cuts the number of levels to divide the full range of cell counts into. This is used for generating the `at` vector. How many different colors are used will depend, to an extent, on this parameter and vice versa. Default is `16`. This is overridden by the length of `at`, if that argument is given.

col.regions numeric vector integers. These correspond to the colors in the graphics device colormap that are to be used to color the hexagons. When `style` is either `"centroids"` or `"lattice"` only the first color is used, so this argument could then be only one number. Different color devices have more or less adequate colormaps. Using a device defined with a call to `trellis.device` will ensure that a colormap close to adequate is used. See the DETAILS section for ways to select other colormaps.

at numeric vector with breaks to determine the mapping from the bin counts to the colormap. Since the colormap has minimal effect on the plot when `style` is either `"centroids"` or `"lattice"`, this argument is not used for those styles. This parameter allows more flexibility in the coloring by allowing non-equally spaced color intervals. If `at` is omitted, it is computed as `pretty(bin$count, cuts)`.

border logical flag: should the border of the hexagon be drawn? (Same as the corresponding argument in the function `polygon`.)

density density of shading lines in lines per inch. If `density` is zero, no shading will occur. If `density` is negative, the hexagons will be filled solidly using the device-dependent polygon filling algorithm. (Same as the corresponding argument in the function `polygon`.) If this argument is given, the user might want to set `border=T` or just the shading lines will appear in the plot.

legend logical flag: should a legend be provided? This implies that space for the legend will be allocated in the plot region.

legend.width numerical value. Width of the space to the right of the plot to be used for the legend in inches. Default is `1`.

legend.lab character string to designate the counts in the legend. Default is `"Counts"`.

legend.cex character expansion parameter for the text in the legend.

xlab x-axis label.

ylab y-axis label.

Graphical parameters may also be supplied as arguments to this function (see `par`).

VALUE

invisibly returns the values of the graphical parameters used for the plot. This list can then be used to add to the plot or to further identify specific points using the function `identify`. See the EXAMPLES below.

SIDE EFFECTS

Produces a plot of the hexagons as determined by the input object. This plot may include a legend or not.

DETAILS

The five plotting styles are:

`style="grayscale"` A smoothly varying color mapping of the counts is determined from the values in `cuts`, `at`, and `col.regions`. The best use of this option requires that the plotting device is activated through a call to the S-PLUS function `trellis.device`. This ensures that an adequate color map is the default although other devices as well as customized colormaps can be provided by the user.

`style="lattice"` or `"centroids"` Plots the hexagons in sizes proportional to cell counts. The `"lattice"` option places the hexagons at the lattice centers. In some cases, the regularity of this structure may be visually overwhelming. In those cases, the user should use the `"centroids"` option which places the hexagons at their centers of mass. This results in the breaking of the regularity of the lattice structure thereby placing the focus on other properties of the data. In all cases the hexagons will not plot outside the cell unless `maxarea > 1`.

`style="nested.lattice"` and `"nested.centroids"` Two overlaying hexagons are plotted: a background hexagon with area equal to the full hexagon's and color proportional to the cell count in *powers of 10* and a foreground hexagon with area proportional to `log10(count)` – `floor(log10(count))`. When `style="nested.centroids"` counts <10 are plotted and the centers of the plotted hexagons are placed at their centers of mass. The outside color encodes hexagon size within color contours representing powers of `10`. Different color schemes give different effects including 3-D illusions.

A way to try different colormaps is by using the *Options* pull down menu in your graphics device driver and typing in your own color map in the Polygons widget. For smoothly varying colormaps, you may want to copy the Image colors into the Polygons slot.

PERCEPTUAL CONSIDERATIONS

Visual response to relative symbol area is not linear and varies from person to person. The argument `at` can be manipulated to give non-linear color variations to aid the interpretation of a hexbin plot. See the EXAMPLES below for some ways to determine an ideal transformation for your colormap.

Plotting the symbols near the center of mass is not only more accurate, but it helps reduce the visual dominance of the lattice structure. Of course higher resolution binning reduces the possible distance between the center of mass for a bin and the bin center. When symbols nearly fill their bin, the plot appears to vibrate. This can be partially controlled by reducing `maxarea` or by reducing contrast.

The local background influences color interpretation. Having defined color breaks to focus attention on specific contours can help. See `nested` options.

WARNING

If `style` is either `"centroids"` or `"lattice"` the legend may consist of overlapping hexagons given the varying areas of these. If that is the case then a smaller value for `cuts` will allow the legend to fit in the space provided for it.

REFERENCES

Carr, D. B. (1991). Looking at large data sets using binned data plots. In *Computing and Graphics in Statistics*. A. Buja and P. Tukey, eds. Springer-Verlag, New York. pp. 7-39.

SEE ALSO

`hexbin, hexagons, identify.hexbin, hex.legend, smooth.hexbin, erode.hexbin`.

EXAMPLES

```
# Simple binning
x <- rnorm(10000)
y <- rnorm(10000)

mybin <- hexbin(x,y)

# Basic plot
plot(mybin,style="nested.lat")

# Lower resolution binning and overplotting with counts
mybin2 <- hexbin(x, y, xbins=20)
screenpar <- plot(mybin2, style="lat", minarea=1, maxarea=1,
                  density=0,border=T)
oldpar <- par(screenpar)  # reset graphic parameters according
                          # to the plot on the screen.
```

```
xy <- cell2xy(mybin2)
text(xy$x, xy$y, format(mybin2$count), adj=.5, cex=.3)

par(oldpar)                 # reset graphic parameters back for
                            # subsequent plotting.

# Histogram equalization to determine amount of color
# Use quantile function to break up the number of counts into
#     more or less equal size groups
at.q <- quantile(mybin$count,probs=seq(10,100,10)/100)
plot(mybin,at=at.q)

# More detail for low counts, use a log transform to determine
# the breaks
at.l <- range(log(mybin$count))
plot(mybin,at=exp(seq(at.l[1],at.l[2],length=10)))
```

`plot.spp`	Plots an Object of class `spp`	`plot.spp`

DESCRIPTION

Plots a spatial point pattern using a scale ratio of 1 for the plot. It allows for overlaying of the object's boundary.

USAGE

```
plot.spp(object, boundary=F, xlim, ylim, ..., bdy.lty=2, bdy.col=1)
```

REQUIRED ARGUMENTS

object a spatial point pattern object. An object of class `"spp"`.

OPTIONAL ARGUMENTS

boundary logical flag: should the object's boundary be added to the plot?

xlim,ylim vectors with 2 components as in `c(min,max)` containing approximate minimum and maximum values to be put on the axes. These values are automatically rounded to make them "pretty" for axis labeling. The range of the data (and the boundary if `boundary=T`) determine their default values.

bdy.lty line type to use when plotting the boundary.

bdy.col color to use when plotting the boundary.

Graphical parameters may also be supplied as arguments to this function (see `par`).

SIDE EFFECTS

an XY-plot of the `object` is produced on the current graphics device.

This function is a method for the generic function `plot` for class `spp`. It can be invoked by calling `plot` for an object of the appropriate class, or directly by calling `plot.spp` regardless of the class of the object.

SEE ALSO

`plot`, `scaled.plot`.

EXAMPLES

```
maples <- spp(lansing[lansing$species=="maple",])
hickories <- spp(lansing[lansing$species=="hickory",])
plot(maples,boundary=T)
```

```
points(hickories)
```

plot.variogram	Plot a Variogram, Covariogram or Correlogram	**plot.variogram**

DESCRIPTION

Display a variogram, covariogram or correlogram object. A multipanel display is created if there is more than one level in the `azimuth` factor in `x`.

USAGE

```
plot.variogram(x, panel=panel.xyplot, ...)
plot.covariogram(x, panel=panel.xyplot, ...)
plot.correlogram(x, panel=panel.xyplot, ...)
```

REQUIRED ARGUMENTS

x   for `plot.variogram`, an object of class `"variogram"`, usually created by the function `variogram`. For `plot.covariogram`, an object of class `"covariogram"`, usually created by the function `covariogram`. For `plot.correlogram`, an object of class `"correlogram"`, usually created by the function `correlogram`.

OPTIONAL ARGUMENTS

panel   a function of two arguments, `x` and `y`, that draws the data display in each panel. The default, `panel.xyplot`, draws points (or lines if `type="l"` is also an argument in the function call).

Graphical parameters may also be supplied as arguments to this function (see `par`). If a multipanel display is to be drawn then the possible graphical parameters are described in the `trellis.args` help file.

SIDE EFFECTS

a plot is produced on the current graphics device.

DETAILS

These functions are methods for the generic function `plot` for objects of classes `"variogram"`, `"covariogram"` and `"correlogram"`. They can be invoked by calling `plot` for an object of the appropriate class.

The axes on the graph are set to include the point `(0,0)` for all plots by default. The function `plot.correlogram` sets the y axis limits to (-1,1).

If the `azimuth` component of `x` has only one level then a standard plot is made of the covariance measure (variogram, covariance or correlation) versus `x$distance`. The function `panel` is still used to draw the points on this standard plot. If there is more than one level in the `azimuth` component then a multipanel display is drawn using the Trellis function `xyplot`. Each panel contains a plot of the covariance measure versus `x$distance` for a particular level of `x$azimuth`.

The default panel function, `panel.xyplot`, uses the `trellis.settings` list for setting `pch`, `col`, `cex` when plotting the points. As such, the plots are best displayed on a device started with the function `trellis.device`.

SEE ALSO

`correlogram`, `panel.xyplot`, `plot`, `trellis.args`, `trellis.device`, `trellis.settings`, `variogram`.

EXAMPLES

```
trellis.device()      # start the default trellis graphics device
vg1 <- variogram(log(tcatch+1) ~ long + lat, data=scallops)
plot(vg1)      # a single panel display
vg4 <- variogram(log(tcatch+1) ~ long + lat, data=scallops,
        azimuth=c(0,45,90,135), tol.azimuth=22.5)
plot(vg4, panel=function(x,y,...) {
        panel.xyplot(x,y,...)
        panel.loess(x,y,...) })      # a 4 panel display with loess smooths
```

plot.vgram.cloud	Plot a Variogram Cloud	plot.vgram.cloud

DESCRIPTION

Display variogram cloud object.

USAGE

```
plot.vgram.cloud(x, ...)
```

REQUIRED ARGUMENTS

x an object of class `"vgram.cloud"`.

Graphical parameters may also be supplied as arguments to this function (see `par`).

SIDE EFFECTS

a plot is produced on the current graphics device.

DETAILS

This function is a method for the generic function `plot` for class `vgram.cloud`. It can be invoked by calling `plot` for an object of the appropriate class, or directly by calling `plot.vgram.cloud` regardless of the class of the object.

SEE ALSO

`variogram.cloud`, `plot`.

EXAMPLES

```
v1 <- variogram.cloud(coal ~ x + y, data=coal.ash)
plot(v1)
```

points.in.poly	Find Points Inside a Given Polygon	points.in.poly

DESCRIPTION

Determine whether points are inside a polygon.

USAGE

```
points.in.poly(x, y, polygon)
```

REQUIRED ARGUMENTS

x the X-coordinates of the points

y   the Y-coordinates of the points. Must be the same length as x.
polygon   a list with named components "x" and "y".

VALUE

a logical vector the same length as x. If TRUE then the corresponding point is inside the given polygon and so on.

BUG

if a ray from a point to an edge intersects a horizontal edge, i.e. is collinear with it, the C program will return TRUE even if such point is not in the polygon.

SEE ALSO

poly.grid, poly.area.

EXAMPLES

points.in.poly(lansing.p$x,lansing.p$y,lansing.chull) # All TRUE

---

poly.area	Computes the Area of a Polygon	**poly.area**

DESCRIPTION

Uses a discrete version of Green's Theorem to calculate the area of a polygon.

USAGE

poly.area(x, y)

REQUIRED ARGUMENTS

x   a list with components named "x" and "y", a 2-column matrix, or a vector containing the horizontal coordinates of the vertices that form the polygon of interest.

OPTIONAL ARGUMENTS

y   if x is a vector of X-coordinates then this must contain the corresponding vertical or Y-coordinates.

VALUE

a double precision number representing the area enclosed by the polygon with vertices defined by x and y

SEE ALSO

points.in.poly, poly.grid

EXAMPLES

```
lansing.chull <- lansing[chull(lansing),]
poly.area(lansing.chull)      # This should be close to unity.
```

`poly.grid`   Generate a Grid Inside a Given Polygonal Boundary   **`poly.grid`**

DESCRIPTION

Generates a grid of points and then clips them to lie within a given boundary.

USAGE

poly.grid(boundary, nx, ny, size)

REQUIRED ARGUMENTS

boundary   a list with components named "x" and "y" or a matrix with 2 columns representing the vertices of a convex polygon. Endpoint need not be repeated.

nx   integer representing the number of cells in the horizontal direction.

ny   integer representing the number of cells in the vertical direction.

OPTIONAL ARGUMENTS

size   numeric vector containing the size of each cell. If it has length one then the cells will be squared with the same side sizes. If it has length two then the cells will have width size[1] and height size[2].

VALUE

a two-column matrix containing the coordinates of the resulting grid.

DETAILS

A rectangular nx by ny grid is overlaid on the polygon defined by boundary and then those points that fall outside are dropped. If size is given then the values nx and ny are redundant and if given will be ignored.

SEE ALSO

points.in.poly

EXAMPLES

```
plot(as.spp(bramble))
bramble.chull <- bramble[chull(bramble),]
polygon(bramble.chull,den=0)
points(poly.grid(bramble.chull,size=c(.1,.8)),col=3)
```

`predict.krige`   Ordinary and Universal Kriging Prediction   **`predict.krige`**

DESCRIPTION

Computes kriging predictions and standard errors at locations in newdata using an object returned by krige.

USAGE

predict.krige(object, newdata, se.fit=T, grid=<<*see below*>>)

REQUIRED ARGUMENTS

object   an object of class "krige" as returned by the function krige.

OPTIONAL ARGUMENTS

newdata   a data frame or list containing the spatial locations for the predictions. The names must match the names of the locations used in the call to krige (see attr(object,"call")).

se.fit a logical value, if TRUE, the standard errors of the predictions are returned. Currently the standard errors are always computed internally. This se.fit only determines if the returned data frame includes the se column.

grid a list containing two vectors, the names of the vectors must match the names of the locations used in the call to krige. The vectors are each of length 3 and specify the minimum, maximum and number of locations in that spatial coordinate, respectively. A grid is then computing using expand.grid. The default value is to use the range of the original location data for the minimum and maximum, and 30 points. This argument is ignored if newdata is supplied.

VALUE

a data frame where the first two columns are the locations of the prediction along with:

fit the predicted values.

se.fit the standard error of the prediction. Only included if se.fit = TRUE.

DETAILS

This function is a method for the generic function predict for class krige . It can be invoked by calling predict for an object of the appropriate class, or directly by calling predict.krige regardless of the class of the object.

REFERENCES

Ripley, Brian D. (1981). *Spatial Statistics.* Wiley, New York.

SEE ALSO

krige, loc.

EXAMPLES

```
# krige the Coal Ash data
kcoal <- krige(coal ~ loc(x, y) + x + x^2, data = coal.ash,
        covfun = spher.cov, range = 4.31, sill = 0.14, nugget = 0.89)
# predictions over default 30 x 30 grid
pcoal <- predict(kcoal)
# plot prediction surface
wireframe(fit ~ x * y, data = pcoal,
        screen = list(z = 300, x = -60, y = 0), drape = T)
```

---

print.slm	Use print() on a slm Object	**print.slm**

USAGE

print.slm(x, ...)

DESCRIPTION

This is a method for the function print() for objects inheriting from class "slm". See print or print.default for the general behavior of this function and for the interpretation of x.

---

print.spatial.cor   Use print() on a spatial.cor Object   **print.spatial.cor**

USAGE

        print.spatial.cor(x, ...)

DESCRIPTION

        This is a method for the function print() for objects inheriting from class "spatial.cor". See print or print.default for the general behavior of this function and for the interpretation of x.

---

print.spatial.neighbor   Print a spatial.neighbor Object   **print.spatial.neighbor**

USAGE

        print.spatial.neighbor(x, ...)

DESCRIPTION

        This is a method for the function print() for objects inheriting from class "spatial.neighbor". See print or print.default for the general behavior of this function and for the interpretation of x.

---

print.summary.slm   Use print() on a summary.slm Object   **print.summary.slm**

USAGE

        print.summary.slm(x, digits, ...)

        This is a method for the function print() for objects inheriting from class "summary.slm". See print or print.default for the general behavior of this function and for the interpretation of x and digits.

---

quad.tree   Order a Multicolumn Real Matrix into a Quad Tree   **quad.tree**

DESCRIPTION

        The quad.tree function performs a recursive partitioning of a numeric matrix, returning row and column index vectors and a list of medians which may be used to "sort" the matrix. The quad tree object is subsequently used in a nearest neighbor search of the matrix (see find.neighbor).

USAGE

        quad.tree(x, bucket.size=1)

REQUIRED ARGUMENTS

        x the numeric matrix from which the quad tree is constructed.

OPTIONAL ARGUMENTS

bucket.size the maximum size for any leaf on the quad tree. This parameter affects the computational time for neighbor search algorithms. The larger the bucket size, the larger the number of observations which need to be examined when performing a search for nearest neighbors on the quad tree. On the other hand, larger bucket sizes may mean that fewer leaves need to be considered. A bucket size of, say, five seems acceptable in many circumstances, though our default of 1 provides a safe value.

VALUE

an object of class `"quad.tree"` containing the following components:

`data` the matrix x.

`nrow` the number of rows in data.

`ncol` the number of columns in data.

`bucket.size` the bucket size used in the quad tree search.

`row.index` the row ordering of the quad tree.

`col.index` the columns used to partition the data into a quad tree.

`medians` the medians used in partitioning the data into a quad tree.

DETAILS

A quad tree is a partitioning of the rows in a matrix which can subsequently be used to efficiently find the rows in the matrix closest (using a variety of metrics) to any given point. Quad trees (also called k-d trees) are thought to be efficient for finding nearest neighbors when the number of columns in the matrix is less than or equal to 10.

The partitioning algorithm proceeds as follows:

1. Set lower = 1 and upper = `nrow` (the index of the first and last row in the matrix x). In the following, only consider the rows of x from lower to upper.
2. Compute the range of each column of x over the range of observations from lower to upper. Set `icol` to the column number with the maximum range.
3. Find the median for column `icol`, and order the rows in the data matrix x over the range lower to upper such that the median evenly splits the rows.
4. If upper - lower = `bucket.size`, return.
5. Go to step 2 with lower unchanged and upper set to (upper+lower)/2 (the left child of the tree).
6. Go to step 2 with lower set to (upper+lower)/2 + 1, and upper unchanged (the right child of the tree).

REFERENCES

Friedman, J., Bentley, J. L., and Finkel, R. A. (1977). An algorithm for finding best matches in logarithmic expected time. *ACM Transaction on Mathematical Software* **3**, 209-226.

SEE ALSO

`find.neighbor`.

EXAMPLES

```
x <- matrix(runif(500),50,10)
quad <- quad.tree(x)

y <- cbind(sids$easting,sids$northing)
sids.quad <- quad.tree(y)
sids.quad <- quad.tree(y, bucket.size=5)
```

`rayplot`	Adds Rays with Optional Confidence Arcs (Sectors)	`rayplot`

DESCRIPTION

Adds rays to an existing plot with angles determined by a third variable and optional confidence sectors around the rays.

USAGE

```
rayplot(x, y, z, lowbnd, highbnd, minz = <<see below>>, maxz = <<see below>>,
   arcs = list(length=0.75, sides=18, lwd=1, density=-1, border=F, col=2),
   tics = list(length=0.06, sides= 8, lwd=1, col=1),
   dots = list(length=0.06, sides= 8, lwd=1, density=0, border=T, col=1),
   rays = list(length=0.30, lwd=2, col=1),
   minangle= -pi/2, maxangle=pi/2, clockwise=F)
```

REQUIRED ARGUMENTS

x,y coordinates of points where the rays are wanted.

z numeric vector of values to translate into the ray angles.

OPTIONAL ARGUMENTS

lowbnd numeric vector. Lower bound of the confidence interval that is desired for z. If provided then it must be the same length as z.

highbnd numeric vector. Upper bound of the confidence interval desired for z. Must be the same length as z, if provided. Either both or none of lowbnd and highbnd must be given.

minz a numeric constant to scale z by. minz maps into minangle so only one of the two is necessary. Its default value is min(z) if lowbnd is missing or min(lowbnd) otherwise.

maxz a numeric constant to scale z by, maxz maps into maxangle so only one of the two is necessary. Its default value is max(z) if highbnd is missing or max(highbnd) otherwise.

Lists of parameters to customize each component of the individual rays can be provided in each of the following arguments:

arcs list of parameters affecting the confidence arcs. This is only relevant if lowbnd and highbnd are provided. The possible components of this list are:

length= numeric scalar or vector giving the confidence sector radii in inches. If a constant, then it is taken to be a fraction of the ray length. If a vector (of the same length as x), then each component is taken to represent the corresponding arc's length.

sides= scalar giving the number of polygon sides for the maximum sector.

lwd= sector line width

density= sector fill options. If density is zero, no filling will occur. If density is negative, the sector will be filled solidly using the device-dependent polygon filling algorithm.

border= logical flag. Should the border of the sector be plotted?

col= integer determining the color for the sector.

tics list of parameters affecting the reference angles against which the user can read the angles of the rays more accurately. Its components are:

length= radius of the tics in inches. No tics are plotted if this is 0 or FALSE.

      sides= number of tics at the base of a ray. Default is 8.

      lwd= tic line width.

      col= integer determining the color of the tics.

dots list of parameters affecting the dot at the base of each ray:

      length= numeric scalar or vector giving the radius of the dot in inches. No dots are plotted if this is 0 or FALSE.

      sides= integer. Number of sides of the "dot."

      lwd= line width parameter.

      density= dot fill parameter. See its definition above for arcs.

      border= logical flag. Should the border of the dot be plotted?

      col= integer determining the color for the dot.

rays The options in the list rays affect the characteristics of the main rays:

      length= numeric scalar or vector giving the ray length in inches. No rays are plotted if this is 0 or FALSE. If this is a scalar, it will be replicated to the a length equivalent to the number of points at which rays are desired, length(x).

      lwd= line width parameter.

      col= integer determining the color of each ray.

minangle numeric scalar giving the minimum angle for the plotting of rays in radians.
maxangle numeric scalar giving the maximum angle for the plotting of rays in radians.
clockwise logical flag, should the direction from the minimum angle to the maximum be clockwise or not?

SIDE EFFECTS

rays with angles from a downwards position are added to the current plot. The smallest value is mapped to -pi/2 and the greatest to pi/2.

DETAILS

Only one direction can be mapped with the current release, if both directions are desired, then call rayplot twice setting clockwise to FALSE and TRUE respectively.

Confidence sectors are useful but should be used with caution. Big arcs will call attention to the less accurate estimates. Consequently, the user may want to use shorter arc lengths for the less accurate estimates to de- emphasize this effect.

The options rays affects the characteristics of the main rays. If the given length for these is a constant value, it will be replicated to the a length equivalent to the number of points at which rays are desired.

The option for tics provides reference angles against which the user can read the ray angles more accurately.

The option for `dots` will put a dot at the base of each ray. Typically, an open dot surrounds the tics.

Outliers are not easily spotted when encoded as ray-angles. If the user wants to highlight selected observations or determine global scaling, `rayplot` should be called for each of the selected data subsets with changing parameters, such as color (`col`), length (`length`), or line width (`lwd`).

REFERENCES

Carr, D. B., Olsen, A. R. and White, D. (1992). Hexagon mosaic maps for display of univariate and bivariate geographical data. *Cartography and Geographics Information Systems* **19**, 228-236.

EXAMPLES

```
# Random locations values and bounds
x <- rnorm(100); y <- rnorm(100); z <- rnorm(100);
inc <- abs(rnorm(100,sd=.4))
lowbnd <- val-inc
highbnd <- val+inc

# No confidence bounds
plot(x,y,type='n')
rayplot(x, y, z)

# Confidence bounds
plot(x,y,type='n')
rayplot(x, y, z, lowbnd, highbnd)

# Clockwise orientation
plot(x,y,type='n')
rayplot(x, y, z, lowbnd, highbnd, clockwise=TRUE)

# No tics and small filled dots
plot(x,y,type='n')
rayplot(x, y, z, lowbnd, highbnd, tic=F,
        dots=list(sides=20, length=.025, density=-1))

# Bivariate rays for smoothed data on a hexagon grid
# (Data not provided)
plot(mymap,type='l')
rayplot(grid$x,grid$y,pred$so4)
rayplot(grid$x,grid$y,pred$no3,clockwise=T,tics=F)
```

---

read.geoeas	Read A GEO-EAS Data File	**read.geoeas**

DESCRIPTION

Read a GEO-EAS data file and create a data frame from it.

USAGE

```
read.geoeas(file)
```

REQUIRED ARGUMENTS

`file` a character string containing the name of the date file to be read.

VALUE

a data frame with as many columns as variables in `file`. The title line and any measurement units listed with the variable names are ignored.

DETAILS

GEO-EAS is a collection of MS-DOS based software tools for performing two-dimensional geostatistical analysis of spatially distributed software. It uses a fixed file format with header information for its data input files. Many other PC based spatial analysis programs also use this file format. See Englund and Sparks (1992) for more details.

REFERENCES

Englund, Evan and Sparks, Allen. (1992). GEO-EAS: Geostatistical Environmental Assessment Software Users's Guide, Version 1.2.1. GWMC-FOS 53 PC, International Ground Water Modeling Center, Golden, CO.

SEE ALSO

read.geoeas, write.table.

EXAMPLES

```
# Read the GEO-EAS file 'iron.geo' into the data frame iron
# (Note the file iron.geo is not included with S+SpatialStats)
iron <- read.geoeas("iron.geo")
```

---

read.neighbor   Read ASCII Files Containing Spatial Contiguity Information   **read.neighbor**

DESCRIPTION

Reads ASCII files containing spatial data with variable length records and creates an object of class "spatial.neighbor".

USAGE

```
read.neighbor(filename, field.names="region.id", region.id=
        length(field.names), first.neighbor=length(field.names)+1,
        keep=T, which.numeric= rep(T, length(field.names)), char=F,
        sep=<<see below>>, skip=0, ...)
```

REQUIRED ARGUMENTS

filename character string giving the directory path and name of an existing file containing the spatial information to be read in. If no path is given, the filename is assumed to be in the current directory. This file is assumed to have one record (row) per spatial unit, which must contain neighborhood (contiguity) information. The number of records per row may vary but some information in addition to the neighbors may appear consistently from record to record. The varying record length is assumed to be completely due to the varying number of neighbors that each region (spatial unit) may have.

OPTIONAL ARGUMENTS

field.names character vector containing the names of the "fixed" fields (in the sense that there is exactly one field per record for all records) in the ASCII file. This vector MUST include the name that we want to give to the region identifier. The default assumes that there are no fields other than the neighborhood information, and hence determines that the only "fixed" field is the region identifier named "region.id".

region.id character string: which of those variables in field.names is to denote the region identifier? This must be an element of field.names. It could also be an integer between 1 and length(field.names) denoting its index in field.names. The default assumes that the "fixed" fields are the laid out first and that the region identifier is the last one of these.

first.neighbor integer denoting the column that contains the first neighbor in the record. It defaults to length(field.names)+1 which implies that the "fixed" fields appear first in each record. This argument allows flexibility on the placement of the varying number of fields (the neighbors) anywhere in the record so long this location is the same from record to record (starting at `first.neighbor`).

keep   logical flag: should we discard or keep the information in the fixed fields? If `TRUE` a data frame is generated with the "fixed" information, if `FALSE` then only the spatial neighborhood information is returned as an object of class `"spatial.neighbor"`. Default is `TRUE`.

which.numeric   a logical vector. Which fixed variables are to be considered numeric? This must be the same length as `field.names`. Its components must be `TRUE` if the corresponding variable in `field.names` is numeric, `FALSE` otherwise. This must be used if any character variables appear in the data or an error will be generated. Default is all numeric fields.

char   logical flag: is the neighborhood information all character? Defaults to `FALSE`, since the most common case is when all neighbors are numeric. If this flag is set to `TRUE`, the resulting object implies a mapping from the character id's onto integers representing the indices of the spatial neighbor matrix.

sep   separator (single character), often `"\t"` for tab or `"\n"` for newline. If omitted, any amount of white space (blanks, tabs, and possibly newlines) can separate fields. Same as argument `sep` for the S-PLUS function `scan`.

skip   the number of initial lines of the file that should be skipped prior to reading. Must be used if the file `filename` contains extraneous information such as a header. Same as argument `skip` for the S-PLUS function `scan`.

...   Routine `scan` is used to read all of the data as character strings. Other arguments to `scan` that can be used include `widths` and `strip.white`. See the help file for this function for more information on these two arguments.

VALUE

if `keep=T` a list with two named components:

data   a data frame containing the variables listed in argument `field.names`.

nhbr   an object of class `"spatial.neighbor"` with the neighborhood information read. Note that this function forces all weights to be equal to 1.

If `keep=F` then an object of class `"spatial.neighbor"` is returned. This is the same as component `nhbr` above.

DETAILS

`read.neighbor` is designed to make the input of spatial information into S+SPATIALSTATS simple for the most common situation.

The file in `filename` is assumed to consist of one record for each region or spatial unit. Each of these records (arranged as rows of `filename`) contains a fixed number of variables, and a varying number of neighboring region identifiers.

As a first step, all data is read as a single vector of character strings. The information provided in arguments `field.names` and `region.id` (and `first.neighbor`, if specified) is then used to convert the data to the resulting data frame and spatial neighbor objects.

If specified, argument `which.numeric` can be used to determine which variables (columns in the returned `data` component) should be of mode numeric or of mode character.

For example, suppose a record contains the following fields:

    city, county, county.id, n1, n2, ..., nk, comments1, comments2

where n1, n2, ... and possibly more are fields containing region identifiers for neighbors of the current region, `county.id` is the field containing an identifier for the current region, and the remaining variables are covariates. In this case, the necessary arguments could be declared as follows:

    field.names = c("city", "county", "county.id", "comments1", "comments2")
    region.id = "county.id"
    first.neighbor = 4
    which.numeric = c(F, F, T, F, F)

and the neighbor fields (n1, n2, ... nk) do not need listing.

SEE ALSO

        `spatial.neighbor, scan, read.table, read.geoeas.`

EXAMPLES

```
# Given the file "test" formatted as follows :
#
#   1 2 3 4
#   2 1 3
#   3 2 1 4
#   4 1 3

read.neighbor("test")

###
# Now if file "test1" is

#    "bill" 1 2 3 4
#    "bob" 2 1 3
#    "mary" 3 2 1 4
#    "pat" 4 1 3

read.neighbor("test1",field.names=c("surveyor","id"),which.numeric=c(F,T))

###
# Another file is "test2"

#    1 2 3 4 "bill"
#    2 1 3 "bob"
#    3 2 1 4 "mary"
#    4 1 3 "pat"

read.neighbor("test2", field.names=c("id","surveyor"), region.id="id", first=2,
            which.numeric=c(T,F))

# Or we may have a different separator and a header, as in file "test3"

#    Neighbourhood information for Survey 5
#    Id Neighbours Surveyor
#    1,2,3,4,"bill"
#    2,1,3,"bob"
#    3,2,1,4,"mary"
#    4,1,3,"pat"

read.neighbor("test3", field.names=c("id","surveyor"), region.id="id",
            first.neighbor=2, which.numeric=c(T,F), sep=",", skip=2)

# Use logical flag char=T if the neighborhood information is character
# For example, say that we have file "test4" with regions A,B,C,D and E.

#    A 23 B C D
#    B 12 A E
#    C 07
#    D 02 B E
#    E 17 A D
```

```
read.neighbor("test4",c("County","Record #"), region.id=1, char=T)

#
read.neighbor("sids.dat", field.names=c("Name","id","nwbirths.ft",
              "sid.ft","births","sid","nwbirths"), region.id="id",
              which.numeric=c(F,T,T,T,T,T,T))
```

---

rfsim	Simulate a Random Field Process	**rfsim**

DESCRIPTION

Simulate an isotropic random field at locations x and y with user supplied covariance function.

USAGE

```
rfsim(x, y, covfun, z=rnorm(length(x)), nc=1000, ...)
```

REQUIRED ARGUMENTS

   x a vector of x locations. Alternatively, x can be a matrix with at least two columns or a list with at least two components. The first column or component contains the x location and the second contains the y locations.

   y a vector of y locations, must be the same length as x.

covfun an S-PLUS function that computes the covariance for the random field as a function of distance. The first argument should be distance. covfun(0, ...) should give the variance of the random field.

OPTIONAL ARGUMENTS

   z a vector of independent, unit variance random values. This must be the same length as x and y. These will be transformed to have covariance covfun.

  nc the number of distance values used in the interpolation of the covariance function. Increasing this value may give a more accurate result at the cost of increased computation time.

  ... additional named arguments that will be passed to covfun, typically the parameters for the covariance function.

VALUE

a vector of random values with covariance function covfun.

SIDE EFFECTS

If z is not supplied, the value of the data set .Random.seed is updated if it already exists, otherwise .Random.seed is created.

DETAILS

The covariance matrix as a function of distance is formed. The z vector is then multiplied by the Cholesky decomposition of this matrix to obtain the result. Linear interpolation over the fixed nh values of distance and cov.fun is used in computing the covariance matrix.

Although the covariance matrix is formed in C code and only the upper triangular is actually formed, it can require large amounts of memory. A 100 x 100 grid of locations will require ((100 * 100) * (100 * 100) + 1) / 2 * 8 bytes / value = 400040000 bytes or about 400 Mb of memory.

SEE ALSO

set.seed.

EXAMPLES

```
# Simulate a Gaussian random field on a 12 x 12 grid with exponential
# covariance function
```

```
x12 <- expand.grid(x=1:12,y=1:12)
z <- rfsim(x12,covfun=exp.cov,range=3)
```

SAR	Simultaneous Spatial Autoregression Object	**SAR**

DESCRIPTION

An object of class `"cov.family"` containing functions for fitting simultaneous autoregression models when used as input to the spatial linear models function, `slm`. See `cov.family.object` for a discussion of the attributes contained in the SAR object. The discussion here centers around the covariance matrix model which the SAR object supports.

DETAILS

Let N denote a neighbor matrix obtained from an object of class `"spatial.neighbor"`, and let Σ denote the covariance matrix of a vector y of spatially correlated dependent variables. Finally, let W denote a diagonal matrix of weights. Then a simultaneous autoregression model assumes that

$$\Sigma = \left[(I - \rho\ N)^T\ W^{-1}\ (I - \rho\ N) \right]^{-1} \sigma^2$$

where ρ is a scalar parameter to be estimated, and σ is a scale parameter which is also to be estimated. This model for the covariance matrix can be generalized to multiple matrices N_i using multiple parameters ρ_i as follows:

$$\Sigma = \left[\sum_i (I - \rho_i N_i) \right]^T W^{-1} \left[\sum_i (I - \rho_i N_i) \right]^{-1} \sigma^2$$

where the N_i are specified through component `matrix` of the `"spatial.neighbor"` object (see routine `spatial.neighbor`). The "regression" aspect of a spatial regression fits the multivariate normal mean vector

$$\mu = E(y|x) = x\beta$$

for unknown parameters β. The multivariate normal likelihood is expressed in terms of the unknowns ρ, σ, and β. The SAR object assumes that a profile likelihood for ρ is fit.

The SAR model can be expressed as an autoregressive model for the spatial parameters as follows:

$$y = X\beta + \rho N(y - X\beta) + W^{1/2}\varepsilon.$$

This allows one to decompose the sum of squares in y into three components (see Haining, 1990, page 258): 1) the trend, $X\beta$; 2) the noise,

$$W^{1/2}\varepsilon = (I - \rho N)(y - X\beta);$$

and 3) the signal,

$$y - X\beta - W^{1/2}\varepsilon.$$

Function `residual.fun` of the SAR object computes ε, the standardized residuals, and routine `slm` returns these in component `residuals`. The estimated trend, $X\beta$, is returned by routine `slm` as the fitted values.

Two functions are required to compute the profile likelihood: 1) a function for computing the determinant $|\Sigma|$, and 2) a function for computing the vector $\Sigma^{-1}z$ for arbitrary vector z. When the single neighbor matrix N is symmetric, the determinant can be expressed and efficiently computed as a function of the eigenvalues of N. If N is not symmetric, or if the dimension of N is large, then sparse matrix routines by Kundert (1988) are used to compute the determinant of Σ. Because the covariance matrix is parameterized in terms of its inverse, the computation of $\Sigma^{-1}z$ is particularly simple and is carried using (sparse) matrix multiplication. See the function `spatial.multiply`.

REFERENCES

Haining, R. (1990). *Spatial Data Analysis in the Social and Environmental Sciences.* Cambridge University Press. Cambridge.

Kundert, Kenneth S. and Sangiovanni-Vincentelli, Alberto (1988). A Sparse Linear Equation Solver. Department of EE and CS, University of California, Berkeley.

SEE ALSO

`cov.family.object, slm, slm.fit, slm.nlminb, spatial.neighbor, MA, CAR, spatial.multiply`.

`scaled.plot`	Equal Scales Plot	`scaled.plot`

DESCRIPTION

Plots a pair of vectors with a given axes scale ratio.

USAGE

`scaled.plot(x, y, scale.ratio=1, xaxs="i", yaxs="i", ...)`

REQUIRED ARGUMENTS

x numeric vector or a list with named components x and y.

y numeric vector. Values to be plotted in the vertical axis.

OPTIONAL ARGUMENTS

scale.ratio numeric value indicating the ratio of unit size in the y-direction to unit size in the x-direction. Default of 1 produces an equally scaled plot.

xaxs,yaxs style of axis interval calculation. See the `par` help file for possible values. The usual default is `xaxs="r"` extends the data range by 4% on each end, and then labels the axis internally. This function sets `xaxs="i"` by default to create an axis labeled internally to the data values. This emphasizes the geometric accuracy of the plot.

Graphical parameters may also be supplied as arguments to this function (see `par`).

SIDE EFFECTS

a plot is created on the current graphics device.

SEE ALSO

`plot`.

EXAMPLES

`scaled.plot(lansing$x, lansing$y, scale=1)`

slm	Fit a Spatial Linear Regression Model	slm

DESCRIPTION

Returns an object of class `"slm"` that represents a fit of a spatial linear (generalized least squares) regression model.

USAGE

```
slm(formula, cov.family, data=<<see below>>, subset=<<see below>>,
    spatial.arglist=NULL, na.action=na.fail, model=F, x=F,
    y=F, contrasts=NULL, ...)
```

REQUIRED ARGUMENTS

formula   a formula object, with the response on the left of a ~ operator, and the terms, separated by + operators, on the right.

cov.family   an object of class `"cov.family"` giving the spatial covariance family to be fit. Valid values are: CAR (conditional auto-regression), SAR (simultaneous auto-regression), or MA (moving average). These are S-PLUS objects containing functions required by the slm fitting algorithm. The covariance model is defined by argument cov.family and is further defined by the variables listed in argument spatial.arglist.

OPTIONAL ARGUMENTS

data   a data frame in which to interpret the variables named in the formula, or in the subset arguments. If this is missing, then the variables in the formula should be on the search list. This may also be a single number to handle some special cases – see NAMES below for details.

subset   this can be a logical vector (with length equal to the number of observations), or a numeric vector indicating which observation numbers are to be included, or a character vector of the row names to be included in the model. All observations are included by default.

spatial.arglist   a list containing arguments required by (and further defining) the spatial model as specified by argument cov.family. Instead of entering these arguments individually, spatial.arglist is used to allow the algorithms to be generalized to different kinds of models. For all of the models currently fit by slm, the spatial.arglist argument contains the following variables:

REQUIRED

neighbor - an object of class `"spatial.neighbor"` containing the neighbors and weights to be used when defining the covariance model (see spatial.neighbor).

OPTIONAL

region.id - a vector containing the rows currently available in the spatial neighbor object. Argument region.id must be given whenever argument subset is given and rows have previously been removed from the spatial neighbor object. This is described below in the DETAILS section. Also see the help file for spatial.subset.

weights - the cov.family uses the neighbor argument to determine a covariance matrix for the residuals. All current types for cov.family allow the specification of a diagonal matrix of weights in the parameterization of the covariance matrix. See the the cov.family for the parameterization. If specified, vector weights contains these diagonal values. If omitted, weights equal to 1 are used.

start - vector of starting values for the optimization algorithm. Since a profile likelihood is optimized, only starting values for the covariance matrix parameters (vector parameters in the output) can be provided. If not provided, these typically default to zero, but this depends upon the cov.family.

print.level - if TRUE, then the function evaluations are printed as the optimization algorithm proceeds. This can be quite useful for checking on convergence of the algorithm to the maximum likelihood estimates.

na.action   a function to filter missing data. This is applied to the model frame after any `subset` argument has been used. The default (with `na.fail`) is to create an error if any missing values are found. A possible alternative is `na.omit`, which deletes observations that contain one or more missing values.

model   logical flag: if `TRUE`, the model frame is returned in component `model`.

x   logical flag: if `TRUE`, the model matrix is returned in component `x`.

y   logical flag: if `TRUE`, the response is returned in component `y`.

qr   logical flag: if `TRUE`, the QR decomposition of the model matrix is returned in component `qr`.

contrasts   a list of contrasts for some or all of the factors appearing in the model formula. Each element of the list should have the same name as the corresponding factor variable, and should be either a contrast matrix (specifically, any full-rank matrix with as many rows as there are levels in the factor), or a function to compute such a matrix given the number of levels.

...   additional arguments which can be passed to the function `slm.nlminb` and which effect the iterative estimation algorithm. In particular, various algorithmic control values can be passed, along with the lower and upper bounds of the parameters.

VALUE

an object of class `"slm"`. Objects of class `"slm"` contain most elements available in class `"lm"` objects (but they do not inherit from class `"lm"` objects), and they also contain items returned by the function `nlminb`. These elements are as follows:

parameters   final values of the parameters over which the optimization takes place. These are the parameters used in defining the covariance structure.

objective   the final value of the objective (-log-likelihood).

message   a statement of the reason for termination.

grad.norm   the final norm of the objective gradient. If there are active bounds, then components corresponding to active bounds are excluded from the norm calculation. If the number of active bounds is equal to the number of parameters, NA will be returned.

iterations   the total number of iterations before termination.

f.evals   the total number of residual evaluations before termination.

g.evals   the total number of jacobian evaluations before termination.

scale   the final value of the scale vector for the minimization.

coefficients   the coefficients of the generalized least-squares fit of the response to the columns of the model matrix. The names of the coefficients are the names of the single-degree-of-freedom effects (the columns of the model matrix). If the model was overdetermined and `singular.ok` was true, there will be missing values in the coefficients corresponding to inestimable coefficients.

residuals   the residuals from the fit. These are not ordinary residuals. See the the `cov.family.object` help file or the `CAR`, `SAR`, or `MA` help files for more information.

fitted.values   the fitted values from the fit. These are the linear trend, $X\%*\%beta$, where X contains the independent variables, and `beta` contains the coefficients of the linear model.

rank   the computed rank (number of linearly independent columns) of the model matrix. If the rank is less than the dimension of R, the columns of R will have been pivoted, and missing values will have been inserted in the coefficients. The upper-left `rank` rows and columns of R are the nonsingular part of the fit, and the remaining columns of the first `rank` rows give the aliasing information (see `alias`).

assign   the list of assignments of coefficients (and effects) to the terms in the model. The names of this list are the names of the terms. The `ith` element of the list is the vector saying which coefficients correspond to the `ith` term. It may be of length 0 if there were no estimable effects for the term.

call   an image of the call that produced the object, but with the arguments all named and with the actual formula included as the formula argument.

contrasts   a list containing sufficient information to construct the contrasts used to fit any factors occurring in the model. The list contains entries that are either matrices or character vectors. When a factor is coded by contrasts, the corresponding contrast matrix is stored in this list. Factors that appear only as dummy variables and variables in the model that are matrices correspond to character vectors in the list. The character vector has the level names for a factor or the column labels for a matrix.

df.residual the number of degrees of freedom for residuals.

model optionally the model frame, if `model=TRUE`.

x optionally the model matrix, if `x=TRUE`.

y optionally the response, if `y=TRUE`.

weights the optional weights (from argument `spatial.arglist`) used in the model.

tau2 the residual variance estimate.

cov.coef the variance-covariance matrix for the coefficients. It is assumed that the estimated coefficients are independent of the covariance matrix parameters (true in the CAR, SAR, and MA models).

DETAILS

slm fits maximum likelihood estimates of spatial regression models (these are equivalent to generalized least squares estimates) using finite difference derivatives and a quasi-Newton optimization algorithm. In such models one assumes a linear model,

$$E(y|x) = x\beta,$$

for the means of the dependent variable given the fixed covariate values, but the errors are assumed to arise from a multivariate normal distribution with a covariance structure as specified by the covariance structure model `cov.family`. See the help files for the MA, CAR and SAR objects for types of covariance structures available (for the usual model based on independent errors, lm may be used.)

The sparse matrix routines of Kundert (1988) are used in solving linear systems and computing determinants required by the likelihood function. The use of these routines makes the algorithm much more efficient than would otherwise be the case. Even so, the cpu time required by the algorithm can be quite large, so lattices with more than, say, 200 to 400 regions should be handled carefully to ensure that cpu time will be available.

A profile likelihood is computed. In this likelihood an equation for the linear model parameters (β) is obtained for known covariance model parameters. Substituting this equation back into the likelihood, the "profile" likelihood is obtained as a function of the covariance model parameters alone. Because a profile likelihood is used, there is a relatively small number of parameters to optimize, making the use of finite difference derivatives more attractive.

Subsetting operations on the spatial data frame are more difficult because the spatial neighbor object must also be subset. This means that a correspondence must be maintained between the "data" object which contains the fixed covariates and the "neighbor" object which maintains information about neighbor relationships. The `region.id` variable of argument `spatial.arglist` provides this correspondence. In the following, for the sake of clarity, we suppose that the linear model is specified via a data frame argument `data`. Vector `region.id` must be the same length as the vectors in the linear model, and the i-th element of `region.id` must "name" the region for the i-th row of `data` in exactly the same manner that the `row.id` and `col.id` values in the `"spatial.neighbor"` object name a region. Then the elements of `region.id` are keys to the `row.id` and `col.id` columns of object `neighbor`. If rows of object `data` are removed, the names of these rows is given by the elements of object `region.id`, and these names are the same names as are used in the `row.id` and `col.id` columns of object `neighbor`. Then rows in `neighbor` can be removed by the subsetting operation.

If the `subset` argument is present, it is evaluated in the context of the data frame, like the terms in `formula`. It is also used in the computation of subsets for any of the arguments contained in `spatial.arglist`, including variables `neighbor` and `region.id`. The specific action of `subset` on the model arguments is as follows: the model frame is computed on *all* rows, then the appropriate subset is extracted. A variety of special cases make such an interpretation desirable (e.g., the use of lag or other functions that may need more than the data used in the fit to be fully defined). On the other hand, if you meant the subset to avoid computing undefined values or to escape warning messages, you may be surprised. For example,

```
slm(y ~ log(x), cov.family = SAR, data = mydata, subset = x > 0)
```

will still generate warnings from `log`. If this is a problem, do the subsetting on the data frame directly:

```
slm(y ~ log(x), cov.family = SAR, data = mydata[mydata$x > 0,])
```

The `subset` argument acts on variable `neighbor` of the `spatial.arglist` argument as follows: Let `region.id` of `spatial.arglist` identify the row numbers in `neighbor` of `spatial.arglist` corresponding to the rows of the data frame given in argument `data`. Then `region.id[subset]` is a listing of the row and column numbers to be used in `neighbor`. Rows and columns of `neighbor` not in the vector `region.id[subset]` are removed.

As in `lm`, the `formula` argument is passed around *unevaluated;* that is, the variables mentioned in the formula in `slm` will be defined when the model frame is computed, not when `slm` is initially called. In particular, if `data` is given, all these names should be defined as variables in that data frame.

Generic functions such as `print` have methods to show the results of the fit.

NAMES. Variables occurring in a formula are evaluated differently from arguments to S-PLUS functions, because the formula is an object that is passed around unevaluated from one function to another. The functions such as `slm` that finally arrange to evaluate the variables in the formula try to establish a context based on the `data` argument. (More precisely, the function `model.frame.default` does the actual evaluation, assuming that its caller behaves in the way described here.) If the `data` argument to `slm` is missing or is an object (typically, a data frame), then the local context for variable names is the frame of the function that called `slm`, or the top-level expression frame if the user called `slm` directly. Names in the formula can refer to variables in the local context as well as global variables or variables in the `data` object.

The `data` argument can also be a number, in which case that number defines the local context. This can arise, for example, if a function is written to call `slm`, perhaps in a loop, but the local context is definitely *not* that function. In this case, the function can set `data` to `sys.parent()`, and the local context will be the next function up the calling stack. See the third example below. A numeric value for `data` can also be supplied if a local context is being explicitly created by a call to `new.frame`. Notice that supplying `data` as a number implies that this is the *only* local context; local variables in any other function will not be available when the model frame is evaluated. This is potentially subtle. Fortunately, it is not something the ordinary user of `slm` needs to worry about. It is relevant for those writing functions that call `slm` or other such model-fitting functions.

REFERENCES

Cliff, A. D. and Ord, J. K. (1981). *Spatial Processes - Models and Applications.* Pion Limited. London.

Cressie, N. A. C. (1993). *Statistics for Spatial Data.* (Revised Edition). Wiley, New York.

Haining, R. (1990). *Spatial Data Analysis in the Social and Environmental Sciences.* Cambridge University Press. Cambridge.

Kundert, Kenneth S. and Sangiovanni-Vincentelli, Alberto (1988). A Sparse Linear Equation Solver. Department of EE and CS, University of California, Berkeley.

Ripley, B. D. (1981). *Spatial Statistics.* Wiley, New York.

There is a vast literature on spatial regression and generalized least squares, the references above are just a small sample of what is available.

SEE ALSO

> CAR, MA, SAR, cov.family.object, lrt.slm, model.frame, slm.fit.spatial, slm.nlminb, spatial.neighbor, spatial.weights.

EXAMPLES

```
sids.maslm <- slm(sid.ft ~ nwbirths.ft, cov.family=MA, data=sids,
    spatial.arglist=list(neighbor=sids.neighbor))

sids.sarslm <- slm(sid.ft ~ nwbirths.ft, cov.family=SAR, data=sids,
    subset=c(-5,-1), spatial.arglist=list(neighbor=sids.neighbor,
    region.id=1:100, weights=1/sids$births))

# myfit calls slm, using the caller to myfit as the local context
# for variables in the formula (see aov for an actual example)
myfit <- function(formula, cov.family, data=sys.parent(), ...) {
    .. ..
    fit <- slm(formula, cov.family, data, ...)
    .. ..
}
```

| **slm.fit** | Fitting Function for Spatial Linear Models | **slm.fit** |

USAGE

> slm.fit(x, y, cov.family, initial, spatial.arglist, ...)

This function is included to be consistent with the lm routines. It performs a few assignments for the slm routine, but these are not critical. See function slm for additional information. For most purposes, you would be better off calling the lower-level function directly unless you want to write code that takes the method argument.

| **slm.fit.spatial** | Fit a Spatial Linear Model (Generalized Least Squares) | **slm.fit.spatial** |

DESCRIPTION

> Fits a spatial linear model (generalized least squares), returning the bare minimum computations.

USAGE

> slm.fit.spatial(x, y, cov.family, initial, spatial.arglist=NULL,
>   start=NULL, ...)

REQUIRED ARGUMENTS

x, y numeric vectors or matrices for the predictors and the response in a linear model. Typically, but not necessarily, x will be the model matrix generated by one of the fitting functions. Note that missing values are not allowed.

cov.family the spatial covariance family to be fit. Valid values are: CAR (conditional auto-regression), SAR (simultaneous auto-regression), or MA (moving average). These are S-PLUS objects containing functions required by the slm fitting algorithm. The covariance model is defined by argument cov.family and is further defined by the variables listed in argument spatial.arglist.

initial a list containing the quantities returned by the cov.family$initialize function, e.g.,

> initial < - cov.family$initialize(spatial.arglist, subset)

spatial.arglist   a list containing arguments required by (and further defining) the spatial model as specified by argument cov.family. Instead of entering these arguments individually, spatial.arglist is used to allow the algorithms to be generalized to different kinds of models. For all of the models currently fit by slm, the spatial.arglist argument contains the following variables:

REQUIRED
neighbor - an object of class "spatial.neighbor" containing the neighbors and weights to be used when defining the model covariance (see spatial.neighbor).

OPTIONAL
region.id - when argument subset is given and rows have been removed from the neighbor object, variable region.id must be used to give the rows currently available in the spatial neighbor object. This is described below in the DETAILS section. Also see the help file for spatial.subset.

weights - the cov.family uses the neighbor argument to determine a covariance matrix for the residuals. It is possible to specify a vector of observation weights to be included in the covariance matrix. Let D denote the diagonal weight matrix, and let S denote the part of the covariance matrix estimate which is based on the neighbor variable in argument spatial.arglist. Then if weights are specified, the residual covariance matrix is computed as matrix expression:
$$D^{1/2} \, S \, D^{1/2}.$$

start - vector of starting values for the optimization algorithm.

print.level - if TRUE, the function evaluations are printed as the optimization algorithm proceeds. This can be useful for checking on convergence of the algorithm to the maximum likelihood estimates.

start   the initial starting value of the parameters, which is a vector with length equal to the number of weight matrices represented by argument neighbor. If it is not specified, the default starting value in spatial.arglist will be used. If that is also not provided nor valid, an initialization procedure will be called to determine the starting value of the parameters.

...   Additional arguments which can be passed to the function slm.nlminb and which effect the iterative estimation algorithm. In particular, various algorithmic control values can be passed, along with the lower and upper bounds of the parameters.

VALUE
A list of class "slm". For a description of the contents of the class, see the function slm.

SEE ALSO
slm, slm.nlminb.

| **slm.nlminb** | Fit a Profile Likelihood in a Spatial Regression Model | **slm.nlminb** |

DESCRIPTION
Optimization algorithm used to estimate spatial regression covariance parameters.

USAGE
```
slm.nlminb(start, objective, scale=1, control=NULL, lower= -Inf,
          upper=Inf, ...)
```

REQUIRED ARGUMENTS
start   p-vector of initial values for the parameters (NAs not allowed).

objective   an S-PLUS function that returns the value of minus the spatial regression profile likelihood function to be minimized. This function must be of the form f(x,<additional arguments>), where x is the vector of parameters over which the minimization takes place. Users can accumulate information through attributes of the value of objective. If the attributes include any additional arguments of objective, the next call to objective will use the new values of those arguments.

OPTIONAL ARGUMENTS

scale   either a single positive value or a positive numeric vector (with length equal to the number of parameters) to be used to scale the parameter vector. Although scale can have a great effect on the performance of the algorithm, it is not known how to choose it optimally. The default is unscaled : scale = 1.

control   a list of parameters by which the user can control various aspects of the minimization. For details, see the help file for nlminb.control.

lower, upper   either a single numeric value or a vector (with length equal to the number of parameters) giving lower or upper bounds for the parameter values. The absence of a bound may be indicated by either NA or NULL, or by -Inf and Inf. The default is unconstrained minimization: lower = -Inf, upper = Inf.

VALUE

a list with the following values:

parameters   final values of the parameters over which the optimization takes place.

objective   the final value of the objective.

message   a statement of the reason for termination.

grad.norm   the final norm of the objective gradient. If there are active bounds, then components corresponding to active bounds are excluded from the norm calculation. If the number of active bounds is equal to the number of parameters, NA will be returned.

iterations   the total number of iterations before termination.

f.evals   the total number of residual evaluations before termination.

g.evals   the total number of jacobian evaluations before termination.

scale   the final value of the scale vector.

aux   the final value of the function attributes.

DETAILS

nlminb is based on the Fortran functions dmnfb, dmngb, and dmnhb. See the nlminb help information for additional details.

Unlike nlminb, slm.nlminb only allows for the specification of the criterion function: finite difference gradients and Hessians are always used. Because a profile likelihood for the spatial regression models is used, there should be very few covariance parameters, and the use of finite difference derivatives will usually not be a problem. In general, the user will not need to call routine slm.nlminb directly.

SEE ALSO

nlminb, slm, slm.fit.spatial.

`smooth.hexbin`	Hexagonal Bin Smoothing	`smooth.hexbin`

DESCRIPTION

Discrete kernel smoother covering seven cells, a center cell and its six neighbors. After two iterations, the kernel covers 1+6+12=19 cells effectively.

USAGE

```
smooth.hexbin(bin, weights=c(48,4,1))
```

REQUIRED ARGUMENTS

`bin` an object of class `"hexbin"`.

OPTIONAL ARGUMENTS

`weights` numeric vector containing the relative weights for the center hexagon, its six first-order neighbor cells, and its twelve second-order neighbors.

VALUE

a data frame with columns `cell` and `count`. Inherits from class `"hexbin"`.

DETAILS

This discrete kernel smoother uses the center cell, immediate neighbors, and second neighbors to smooth the counts. The counts for each resulting cell are a linear combination of neighboring cell counts and weights.

If a cell, its immediate, and its second neighbors all have a value of `max(bin$count)` then the new maximum count is `max(bin$count)*sum(weights)*(1+6+12)`.

Set `weights[3]=0` if only immediate neighbors' effects are desired.

The current implementation increases the domain `attr(bin,"dims")` by four rows and four columns, thus reducing plotting resolution.

REFERENCES

Carr, D. B. (1991). Looking at large data sets using binned data plots. In *Computing and Graphics in Statistics*. A. Buja and P. Tukey, eds. Springer-Verlag, New York. pp. 7-39.

SEE ALSO

`hexbin`, `plot.hexbin`, `erode.hexbin`.

EXAMPLES

```
# Show the smooth counts in gray level
smoothbin  <- smooth.hexbin(mybin)
plot(smoothbin)

# Compare the smooth and the origin
smbin1 <- smbin
smbin1$count <- ceiling(smbin$count/sum(smbin$weights))
plot.hexbin(smbin1)
# Expand the domain for comparability
smbin2 <- smooth.hexbin(bin, weights=c(1,0,0))
plot.hexbin(smbin2)
```

`spatial.cg.solve`	Solve S b = x	`spatial.cg.solve`

DESCRIPTION

Solves S b = x for b, where S is a sparse matrix obtained from an object of class `"spatial.neigh-bor"`.

USAGE

```
spatial.cg.solve(neighbor, x, transpose=F, print.level=F, rho=0,
                 product=F, region.id=NULL)
```

REQUIRED ARGUMENTS

neighbor   an object of class `"spatial.neighbor"` containing the sparse matrix representation of the spatial neighbor matrix (or matrices, see function `spatial.neighbor`).

x   the right hand side for which a solution is desired. Alternatively, x can be a matrix. In this case, a solution is obtained for each column in x.

OPTIONAL ARGUMENTS

transpose   with the default arguments, S is taken as I minus the sum over i of rho[i] A[i]. Here I is an identity matrix, rho[i] is a scalar, and A[i] is the i-th weight matrix in neighbor. If transpose is T, then the transpose of this matrix is used for A.

print.level   if TRUE, information regarding the iterative process is printed as the iterations are performed. This allows you to better judge the adequacy of the solution.

rho   a scalar (or vector) of constants used in defining the matrix S (see argument transpose).

product   let B be I - sum rho[i] A[i] as described in argument transpose. When product is FALSE, S = B. When product is TRUE, S is t(B) times B.

region.id   a vector with length equal to the number of regions in the spatial lattice. If variables row.id and col.id of argument neighbor are not integer valued variables with sequential values from 1 to the number or regions in the lattice, then argument region.id must be specified and is used to obtain a sequential coding of the lattice regions.

VALUE

a matrix (or vector), b, solving the linear system S b = x.

DETAILS

A iterative algorithm due to M. A. Saunders (see Paige and Saunders, 1975) is used to solve the linear system. This algorithm requires the matrix products S y, for vector y. In `spatial.cg.solve` this product is computed using the same algorithms as in routine `spatial.multiply`. The Paige and Saunders algorithm is a variant of a conjugate gradient algorithm.

REFERENCES

Lewis, J. G.(1977). "Algorithms for Sparse Matrix Eigenvalue Problems". Report STAN-CS-77-595, Computer Science Department, Stanford University, Stanford, California

Paige, C. C. and Saunders, M. A. (1975). Solution of sparse indefinite systems of linear equations. *SIAM J. Numer. Anal.* **12**, 617-629.

SEE ALSO

`spatial.neighbor`, `spatial.multiply`.

EXAMPLES

```
row.id <- c(1,1,2,2,3)
col.id <- c(1,3,1,3,4)
x <- 1:4
```

```
neighbor <- spatial.neighbor(row.id=row.id, col.id=col.id, symmetric=T)
spatial.cg.solve(neighbor, x, rho= -0.7)
```

---

spatial.condense   Remove Redundancy in `"spatial.neighbor"` Objects   **spatial.condense**

DESCRIPTION

Eliminates redundancy in an object of class `"spatial.neighbor"` which may exist if the neighbor relationships are known to be symmetric.

USAGE

```
spatial.condense(neighbor, symmetric=attr(neighbor, "symmetric"))
```

REQUIRED ARGUMENTS

`neighbor` an object of class `"spatial.neighbor"` containing the sparse matrix representation of the spatial neighbor matrix (or matrices). See function `spatial.neighbor`.

OPTIONAL ARGUMENTS

`symmetric` if `TRUE`, asymmetric terms in the neighbor object are removed and the symmetric attribute of the resulting object is reset to `TRUE`.

VALUE

an object of class `"spatial.neighbor"` representing the same sparse matrix as the input object, but with redundant entries removed. No attempt at combining entries is made. If an entry is determined to be redundant, it is removed.

DETAILS

Argument `neighbor` is a sparse matrix representation of one or more matrices. It must be an object of class `"spatial.neighbor"`.

Let `a[i,j]` denote an element of one such matrix. Then if `a[i,j]` is not zero, regions `i` and `j` are neighbors, and the strength of the relationship is given by the magnitude of `a[i,j]`.

If the object of class `"spatial.neighbor"` is labeled symmetric, then only one of `a[i,j]` or `a[j,i]` is needed to specify it.

In the case of the object having more than one neighbor matrices (i.e. attribute `nregion>1`), the k-th matrix may be combined with the l-th matrix. It is again possible that two different values for `a[i,j]` might be defined. Routine `spatial.condense` eliminates the redundant values by removing all but one reference to each neighbor pair from all matrices corresponding to the spatial neighbor object.

SEE ALSO

`spatial.neighbor`, `spatial.weights`.

EXAMPLES

```
index.rows <- c(1,1,2,2,3)
index.cols <- c(2,3,1,3,4)
my.neighbor <- spatial.neighbor(row.id=index.rows, col.id=index.cols)
attr(my.neighbor, "symmetric") <- T    # The default had set it to FALSE
print(my.neighbor)
print(spatial.weights(my.neighbor))
my.neighbor <- spatial.condense(my.neighbor)
print(my.neighbor)
print(spatial.weights(my.neighbor))
```

spatial.cor	Measures of Spatial Correlation	**spatial.cor**

DESCRIPTION

Computes the Moran, Geary, and other measures of spatial correlation. Also computes a Monte Carlo estimate of the distribution of the spatial correlation statistic.

USAGE

```
spatial.cor(x, neighbor, statistic="moran", sampling="nonfree",
        npermutes=0, weight.fun=NULL, cov.fun=NULL)
```

REQUIRED ARGUMENTS

x   numeric vector or matrix containing the spatial observations. If a matrix, then each row in the matrix (or each element in a vector) corresponds to a different spatial region.

neighbor   an object of class "spatial.neighbor" containing the spatial weights and specifying the spatial connectivity matrix. It is assumed that observations from connected spatial regions are correlated, and that the strength of their correlation is given by the spatial weights specified in the spatial neighbor object.

OPTIONAL ARGUMENTS

statistic   a character string to select the statistic to be used in computing the spatial correlation measure. This can be one of "moran", "geary" or "user". Partial matching is allowed.

The choices are:

"moran" - the Moran (1950) measure of spatial association.

"geary" - the Geary (1954) index of spatial association.

"user" - a user specified measure of spatial association.

The statistic="user" option allows the user to define their own correlation measure. In this case, estimates of the variance cannot be computed, though it is still possible to compute a Monte Carlo estimate of the permutation distribution. When statistic="user", the arguments weight.fun and cov.fun must be specified.

sampling   a character string giving the sampling assumptions to be used when computing the variances. Two sampling assumptions are possible: "nonfree", in which the variance, conditioned upon the observed values of argument x, is computed; and "free", in which each of the observations in argument x is free to vary. See Cliff and Ord (1981, page 12) for discussion. Variances are not computed if statistic="user". Partial matching is allowed.

npermutes   integer value. The number of permutations to be used in estimating the distribution of the spatial correlation measure. The permutation distribution estimate is computed as follows: npermutes random permutations of the (rows of) data in x are generated, and for each permutation the spatial correlation measure is computed for each column in x. The vector of permutation correlations provide an estimate of the permutation distribution of the spatial correlation for each column in x. The permutation distribution is the distribution one would obtain if a random permutation of the data were used to compute the spatial correlation measure. That is, the permutation distribution is the null distribution of the test statistic, conditional upon the observed data values.

As with many Monte Carlo simulations, npermutes=100 is often satisfactory for estimating p-values, though additional precision is obtained when more permutations are taken. More observations are commonly used when confidence intervals are to be computed. See, e.g., Good (1994, page 163), for a discussion.

weight.fun when argument `statistic="user"`, you must supply an S-PLUS weight function. This weight function must have two arguments, `(x,A)`, where `x` is a vector representing a single column of the input matrix `x`, and `A` is the sum of the weights. Using these two arguments, a normalizing constant is computed, and the spatial correlation measure is computed as the product of this normalizing constant times the measure of association computed by the function given in the argument `cov.fun`. For the Moran measure of spatial correlation, the weight function is:

```
MORAN.weight <- function(x, A) {
    length(x)/(A * (length(x) - 1) * var(x))
}
```

cov.fun when argument `statistic="user"`, you must provide an S-PLUS function for computing the covariances in your correlation measure. The covariance function has four arguments, `(x, row.id, col.id, weights)`, where `x` is a vector containing (a single column of) input matrix `x`, and `row.id`, `col.id`, and `weights` are vectors specified in argument `neighbor` giving the connections between observations in `x` (see function `spatial.neighbor`). Observations `row.id[i]` and `col.id[i]` in `x` are connected with weight given by `weights[i]`. For example, the Moran covariance function is:

```
MORAN.cov <- function(x, row.id, col.id, weights) {
    m <- mean(x)
    sum(weights * (x[row.id] - m) * (x[col.id] - m))
}
```

VALUE

an object of class `"spatial.cor"` with components:

statistic the statistic used in computing the correlation measure. Same as its input value.

sampling the sampling assumption used for variance estimation. Same as its input value.

n the number of observation or sampling units.

correlation the spatial correlation estimate.

variance estimate of the variance of the spatial correlation estimates. Variances are not computed if `statistic="user"`.

perm.p.value one-sided p-value computed using the permutation distribution. Each p-value gives the probability (as estimated from the permutation distribution) that the corresponding spatial correlation measure is larger than the observed value, under the null hypothesis of no spatial correlation.

perm.corr permutation estimates of the correlation measures. Confidence intervals and other quantities of interest can be computed from the (permutation) distribution of these estimates.

The print method, `print.spatial.cor`, prints out the normal z statistic and its two-sided p-value for the null hypothesis of no spatial correlation when `statistic` is `"moran"` or `"geary"`.

DETAILS

The routine `spatial.cor` computes one of two built-in measures of spatial correlation. These are:

The Moran coefficient:

$$M = \frac{n}{A} \frac{\sum_{i,j} w_{i,j}\, z_i\, z_j}{\sum_i z_i^2}$$

Here $w_{i,j}$ is the weight for the relationship between observations `i` and `j` (zero means no relationship), $A = \sum w_{i,j}$, and $z_i = x_i - \bar{x}$ is the centered variate obtained from x_i.

The Geary coefficient:

$$G = \frac{(n-1)}{2A} \frac{\sum_{i,j} w_{i,j}(x_i - x_j)^2}{\sum z_i^2}$$

The measures (except those for `statistic="user"`) are described in Cliff and Ord (1981, Chapter 1).

The Moran measure most resembles a Pearson correlation coefficient, and has mean $-1/(n-1)$ when there is no association. Here n is the number of rows in x (or, for vectors, n is the length of x). The Geary measure has mean 1 in the null case.

In addition to these two measures, you can specify other measures of spatial association by providing an S-PLUS function to compute a weighting or scaling factor (the `weight.fun` function), with a second function to compute a covariance or association measure (the `cov.fun` function). These functions, whose arguments are obtained from the input arguments to `spatial.cor`, each return a single value, and the measure of association is computed as the product of these values. Routine `call_S` is used in computing the permutation distribution for user specified correlations. `call_S` is somewhat slower (and uses more memory) than the C code used in computing the permutation distribution for the built-in measures.

Permutation distributions are important when computing measures of spatial correlation because the null distribution of the association statistic varies with the spatial lattice size and shape. This variability makes it difficult to provide approximate theoretical distributions, making the distribution of the Monte Carlo estimates all the more valuable. Confidence intervals and tests can be computed from the permutation distribution as they would be from an exact distribution. For example, a two-sided 10 percent confidence is obtained as the 5-th and 95-th percentile from the permutation distribution. Notice, however, that different runs of the program with different random number seeds will lead to slightly different results. Use `set.seed` to set the random number seed.

REFERENCES

Cliff, A. D. and Ord, J. K. (1981). *Spatial Processes: Models and Applications.* Pion Limited, London.

Geary, R. C. (1954). The contiguity ratio and statistical mapping. *The Incorporated Statistician* **5**, 115-145.

Good, P. (1994). *Permutation Tests.* New York. Springer Verlag.

Moran, P. A. P. (1948). The interpretation of statistical maps. *Journal of the Royal Statistical Society, Series B* **10**, 243-251.

Moran, P. A. P. (1950). Notes on continuous stochastic phenomena. *Biometrika* **37**, 17-23.

SEE ALSO

`spatial.weights, spatial.neighbor, call_S, set.seed`

EXAMPLES

```
sids.cor <- spatial.cor(sids$sid, neighbor=sids.neighbor, statistic="geary",
      sampling="free", npermutes=100)
sids.cor
```

| **`spatial.determinant`** | Compute Sparse Matrix Determinant | **`spatial.determinant`** |

DESCRIPTION

Function to compute the determinant of $I - A$ where $A = \sum_i \rho_i N_i$, N_i is the i-th spatial neighbor matrix, and ρ_i is an element of the parameter vector `parameters`.

USAGE

```
spatial.determinant(neighbor, parameters=NULL, absThreshold=0,
                 relThreshold=0, diagPivoting=0, shareMemory=F)
```

REQUIRED ARGUMENTS

neighbor an object of class `spatial.neighbor` containing the sparse matrix representation of the spatial neighbor matrices `N[i]` (see function `spatial.neighbor`).

OPTIONAL ARGUMENTS

parameters a vector of scalars multiplying the neighbor matrices. If specified, the length of the vector must be at least the number of neighbor matrices defined in the neighbor object (= `length(unique(neighbor$matrix))`)

absThreshold this is the pivot threshold, which should be between zero and one. If it is one then the pivoting method becomes complete pivoting, which is very slow and tends to fill up the matrix. If it is set close to zero the pivoting method becomes strict Markowitz with no threshold.

relThreshold the absolute magnitude an element must have to be considered as a pivot candidate, except as a last resort.

diagPivoting if `TRUE`, pivot selection should be confined to the diagonal if possible.

shareMemory a logical flag indicating if the memory will be shared by other routines. If `TRUE`, memory is shared and it needs to be released later. One way to release the memory is call `.C("destroy_sparse_matrix")` after the shared memory is no long needed.

VALUE

a list containing two named variables:

mantissa the mantissa of the calculated determinant.

exponent the exponent of the calculated determinant (base 10).

DETAILS

The neighbor object is a sparse matrix representation of the neighbor matrix. Because it is sparse, the matrix determinant calculation is carried out using sparse matrix methods.

REFERENCES

Kundert, Kenneth S. and Sangiovanni-Vincentelli, Alberto (1988). A Sparse Linear Equation Solver. Department of EE and CS, University of California, Berkeley.

SEE ALSO

`spatial.cg.solve`, `spatial.multiply`, `spatial.neighbor`, `spatial.neighbor.object`.

EXAMPLES

```
spatial.determinant(sids.neighbor, parameters=0.01)
```

`spatial.multiply`   Compute Sparse Matrix Vector Product A x   **`spatial.multiply`**

DESCRIPTION

Function to compute product `Ax` where $A = \sum_i \rho_i N_i$, where N_i is the i-th spatial neighbor matrix, ρ_i is an element of the parameter vector `parameters`, and `x` is an arbitrary vector or matrix.

USAGE

```
spatial.multiply(neighbor, x, transpose=F, parameters=NULL,
                 region.id=NULL)
```

REQUIRED ARGUMENTS

neighbor an object of class `spatial.neighbor` containing the sparse matrix representation of the spatial neighbor matrices N_i (see function `spatial.neighbor`).

x a vector or matrix for which the product is desired.

OPTIONAL ARGUMENTS

transpose a logical value indicating whether the transpose of the spatial neighbor matrices N_i are to be used in place of the N_i when computing `A`. Here `A` is represented as $\sum_i parameters_i N_i$, where $parameters_i$ is a scalar, and N_i is the i-th spatial neighbor matrix represented by `neighbor`. If `transpose` is `TRUE`, then the transpose of each N_i is used in place of N_i.

parameters a vector of scalars multiplying the neighbor matrices.

region.id a vector with length equal to the number of regions in the spatial lattice. If variables `row.id` and `col.id` of argument `neighbor` are not integer valued variables with sequential values from 1 to the number or regions in the lattice, then argument `region.id` must be specified and is used to obtain a sequential coding of the lattice regions.

VALUE

a vector (or matrix) containing the product `A x`.

DETAILS

The neighbor object is a sparse matrix representation of the neighbor matrix. Because it is sparse, the matrix multiplication is carried out using sparse matrix methods.

SEE ALSO

`spatial.cg.solve`, `spatial.neighbor`, `spatial.neighbor.object`.

EXAMPLES

```
spatial.multiply(sids.neighbor, 1:100, parameters=c(-0.3))
```

`spatial.neighbor`   Create a `"spatial.neighbor"` Object   **`spatial.neighbor`**

DESCRIPTION

Function used to create an object of class `"spatial.neighbor"` given its component parts.

USAGE

```
spatial.neighbor(neighbor.matrix, nregion=dim(neighbor.matrix)[1])
spatial.neighbor(row.id, col.id, weights=rep(1, length(row.id)),
                 nregion=max(c(row.id,col.id)), symmetric=F,
                 matrix.id=<<see below>>)
```

REQUIRED ARGUMENTS

neighbor.matrix   a matrix of neighbor weights (where all weights are often 1) from which the object of class "spatial.neighbor" is to be constructed. This must be a square matrix such that if element [i,j] is non-zero, then spatial regions i and j are considered neighbors, and its value is used as a weight in measures of correlation or in further model-fitting. This is also known as the contiguity matrix.

row.id   integer vector. The row indices of the non-zero elements of the neighbor weight matrix. The i-th element of row.id and the i-th element of col.id specify two regions which are spatial neighbors. Two regions are spatial neighbors if observations from the two regions have a non-zero spatial weight and vice-versa. This is ignored if neighbor.matrix is given.

col.id   integer vector (of the same length as row.id) with the column indices of the non-zero elements of the neighbor weight matrix. This is ignored if neighbor.matrix is given.

It is important to note that even if a pair of regions c(row.id[i],col.id[i]) are spatial neighbors, the permuted pair c(col.id[i],row.id[i]) does not have to define spatial neighbors (corresponding contiguity matrix element can be zero). For example, consider two regions on a river, and suppose that a region corresponding to row.id[i] is downstream from the region in col.id[i] and neighbors. By this definition, "downstream of" the transpose pairing need not satisfy a neighbor relationship. See argument symmetric below.

OPTIONAL ARGUMENTS

weights   numeric vector of the same length as row.id and col.id. weights[i] gives a weight for the corresponding neighbor pair relationship, given in c(row.id[i],col.id[i]). If weights is not specified (and argument neighbor.matrix is not used), then the spatial weights are all set equal to 1. Each spatial weight defines the strength of the association between two neighbors. This argument is ignored if neighbor.matrix is given as each of the matrix elements are then considered to be neighbor weights.

nregion   integer stating the total number of regions or spatial units. If not given, this value is computed from the number of unique elements in row.id and col.id as the maximum of all the regions given therein max(c(row.id,col.id)).

symmetric   logical flag: should the neighbor matrix be considered symmetric?. If TRUE, the spatial weights matrix is computed by assuming that if the i-th neighbor pair c(row.id[i],col.id[i]) has neighbor weight given by w=weights[i] then so does the matrix element c(col.id[i],row.id[i]). Only half of the weights need be specified in this case. If TRUE, routine spatial.condense is called to remove redundant values. When neighbor.matrix is given, its symmetry is determined within the function, otherwise it defaults to FALSE.

matrix.id   integer vector of length equal to the total number of spatial neighbors. This can be used to differentiate various types of neighbors. For example, spatial regression models may differentiate between north-south neighbors as compared to east-west neighbors. The values of vector matrix.id should then indicate the neighbor types. If missing, a single neighbor type is assumed (with one neighbor matrix).

VALUE

an object of class "spatial.neighbor". This object inherits from class "data.frame" and describes the relationship among spatial regions using a sparse representation of the Weight or Contiguity matrix (or matrices). It has columns row.id, col.id, weights and matrix (determined by matrix.id).

DETAILS

Objects of class "spatial.neighbor" are required by the spatial regression, spatial correlation, and other functions in S+SPATIALSTATS. Two methods for constructing a spatial neighbor object are available. A matrix of weights (where all weights are often 1) can be given as input, and the "spatial.neighbor" object is constructed from its non-negative elements. In this case argument neighbor.matrix must be a square matrix such that if element c(i,j) of the matrix is non-zero, then spatial regions i and j are neighbors, with weight given by the value of the element (usually a 1).

Another method for constructing an object of class `"spatial.neighbor"` is by directly specifying the row and column numbers (and the weight value) of the non-zero elements of the contiguity matrix which is usually a sparse matrix. A sparse representation is usually preferred in practice. In this case, `row.id[i]` gives the row of the i-th non-negative element of the neighbor matrix, and the corresponding element `col.id[i]` gives its column index. Thus, each pair `c(row.id[i],col.id[i])` represents a pair of neighboring spatial units. The strength of their association can then be given by `weights[i]`.

Notice that `row.id` and `col.id` contain INDICES of the contiguity matrix and NOT the region identifiers which could be character strings or some such. These are used to expand the full contiguity matrix, so we should have representation for all indices 1 through `nregion`, though it is possible to have islands in between. Use the function `check.islands` to check for these islands, and remap their indexing if that is desirable.

It is possible to specify two or more types of neighbor relationships. For example, the user may want to model a spatial relationship depending upon the angle of the line connecting neighbor centers i.e. considering directional relationships. For this example, let Type-1 neighbors be north-south neighbors, and let Type-2 neighbors be east-west neighbors; neighbors along a diagonal could be modeled with weights proportional to `.707` (the sine of 45 degrees), for instance.

Consider the elements of `row.id`, `col.id`, and `weights` corresponding to a distinct value, k, of the vector `matrix.id`. The spatial neighbor matrix can be expressed as a matrix `A[k]` such that `A[k][row.id,col.id]=weights`, and all other elements are zero. Consider a parameter vector `rho` of length g, many spatial covariance matrices used in spatial regression models can be expressed as a weighted linear combination of the contiguity matrices `A[k]`, `rho[k]*A[k]`, for values of k varying in `1:g`.

SEE ALSO

is.Hermitian, check.islands.

EXAMPLES

```
row.index <- c(1,1,2,2,3)
col.index <- c(2,3,1,3,4)
# Assume we have no information about the strength of the spatial
# association.  All weights are 1.
neighbour <- spatial.neighbor(row.id=row.index, col.id=col.index)
```

---

spatial.neighbor.object   Class `"spatial.neighbor"`   **spatial.neighbor.object**

DESCRIPTION

Class of objects used to define neighbor relationships for spatial data on a regular or irregular lattice.

GENERATION

This class of objects is constructed using the function `spatial.neighbor`. Alternatively, the functions `read.neighbor`, or `neighbor.grid` may be used. In general, the user must construct these objects whenever estimates of spatial correlation and spatial regression are desired.

An object of class `"spatial.neighbor"` contains all the information required to determine which spatial units on a region of interest are neighbors, as well as the strength of their relationship.

METHODS

The class `"spatial.neighbor"` has one associated method, `print.spatial.neighbor`.

INHERITANCE

Class `"spatial.neighbor"` inherits from class `"data.frame"`.

STRUCTURE

The `"spatial.neighbor"` object is in essence a data frame with additional attributes. Each row of the data frame denotes a pair of neighboring spatial units. The data frame contains the following columns:

`row.id` the row index in the neighbor matrix that corresponds to a region or spatial unit. This implies a numbering of regions from 1 to the total number of regions.

`col.id` the column index in the neighbor matrix that corresponds to the neighbor of the region defined by the corresponding element of `row.id`.

`weights` a numeric value giving the relative strength of the neighbor relationship. The larger the value, the stronger the relationship.

`matrix` if multiple types of neighbor matrices are possible, this column contains the type of the neighbor this weight represents - it gives a numeric identifier for each spatial neighbor [contiguity] matrix.

SPECIAL ATTRIBUTES

`nregion` the number of total regions in the study. The row and column identifiers given in `row.id` and `col.id` might not include ALL the spatial units in the area of interest. This happens when units are isolated, i.e. have no neighboring regions. In this case, `nregion` must be used to determine the total number of rows and columns in the contiguity matrix.

`symmetric` a logical flag; if TRUE, the contiguity matrix is assumed to be symmetric. If TRUE, only the weights for the upper (or lower) triangle of the contiguity matrix need to be specified in the object. Use the function `spatial.weights` to expand the full symmetric weights matrix.

DETAILS

An object of class `"spatial.neighbor"` is a sparse matrix representation of a square matrix (or a number of square matrices).

The functions `spatial.multiply`, and `spatial.cg.solve` can be used to form products of the form `rho[i]*N[i]*x` and `(rho[i]*N[i])^(-1)*x`, for neighbor weight matrices `N[i]`, vector of constants or parameters, `rho[i]`, and arbitrary vectors `x`, should that be needed to form a neighbor or contiguity matrix as a weighted linear combination of others.

SEE ALSO

`spatial.neighbor`, `read.neighbor`, `neighbor.grid`, `spatial.multiply`, `spatial.cg.solve`, `spatial.weights`.

`spatial.solve`	Solve S b = x	`spatial.solve`

DESCRIPTION

Solves S b = x for b, where S is a sparse matrix obtained from an object of class `"spatial.neighbor"`.

USAGE

```
spatial.solve(neighbor, x, transpose=F, rho=0, product=F,
              weights=NULL, region.id=NULL, absThreshold=0,
              relThreshold=0, diagPivoting=0, shareMemory=F)
```

REQUIRED ARGUMENTS

`neighbor` an object of class `"spatial.neighbor"` containing the sparse matrix representation of the spatial neighbor matrix (or matrices, see function `spatial.neighbor`).

x the right hand side for which a solution is desired. Alternatively, x can be a matrix. In this case, a solution is obtained for each column in x.

OPTIONAL ARGUMENTS

transpose with the default arguments, S is taken as I minus the sum over i of rho[i] * A[i]. Here I is an identity matrix, rho[i] is a scalar, and A[i] is the i-th weight matrix in neighbor. If transpose is TRUE, then the transpose of this matrix is used for A.

rho a scalar (or vector) of constants used in defining the matrix S (see argument transpose).

product let B=I minus the sum of rho[i]*A[i] as described in argument transpose. When product is FALSE, S=B. When product is TRUE, S is t(B)%*%B.

weights if provided, the inverse weights are included along the diagonal matrix W and incorporated into the model for S as follows: Let R be I minus the sum of rho[i]*A[i]. Then

```
product | transpose |   S
---------------------------------
  F     |    F      | R %*% W
  F     |    T      | t(R) %*% W
  T     |    F      | t(R) %*% W %*% R
  T     |    T      | R %*% W %*% t(R)
```

region.id a vector with length equal to the number of regions in the spatial lattice. If variables row.id and col.id of argument neighbor are not integer valued variables with sequential values from 1 to the number or regions in the lattice, then argument region.id must be specified and is used to obtain a sequential coding of the lattice regions.

absThreshold the pivot threshold (between zero and 1). Values near 1 result in complete pivoting, while values near zero result in a strict Markowitz solution. In general, you should choose a value as close to zero as roundoff error will permit. A value of 0.001 has been recommended by Kundert (1988) in some cases.

relThreshold the absolute magnitude an element must have to be considered as a pivot candidate, except as a last resort. This should be set to a small fraction of the smallest (absolute) diagonal element.

diagPivoting if TRUE, pivot selection should be confined to the diagonal if possible.

shareMemory if TRUE, the in-memory representation of the sparse matrix will be shared by other routines. If memory is shared, it needs to be released later. One way to release the memory is to call .C("destroy_sparse_matrix") after the in-memory representation of the matrix is no long needed. Most users should use the default value, FALSE.

VALUE

a matrix (or vector), b, solving the linear system S b = x.

DETAILS

This routine uses the sparse matrix code of Kenneth Kundert and Alberto Sangiovanni-Vincentelli (1988). The University of California, Berkeley, holds the copyright for these routines.

REFERENCES

Kundert, Kenneth S. and Sangiovanni-Vincentelli, Alberto (1988). A Sparse Linear Equation Solver. Department of EE and CS, University of California, Berkeley.

SEE ALSO

spatial.cg.solve, spatial.multiply, spatial.neighbor, spatial.neighbor.object.

EXAMPLES

```
x <- 1:4
row.id <- c(1,1,2,2,3)
col.id <- c(1,3,1,3,4)
alpha <- 0.3
neighbor <- spatial.neighbor(row.id=row.id, col.id=col.id, symmetric=T)
a <- solve(diag(attr(neighbor, "nregion"))-alpha*
        spatial.weights(neighbor), x)
```

```
b <- spatial.solve(neighbor, x, rho=alpha)$result
print(max(abs(a-b)) < 1e-14)
```

---

spatial.subset	Subset an Object of Class `"spatial.neighbor"`	**spatial.subset**

DESCRIPTION

Obtains a subset of an object of class `"spatial.neighbor"`.

USAGE

```
spatial.subset(neighbor, region.id, data=NULL, subset=NULL,
               reorder=F, parameters=NULL, weights=NULL)
```

REQUIRED ARGUMENTS

neighbor an object of class `"spatial.neighbor"`.

region.id a vector of id's for the set of all spatial units. The `row.id` and `col.id` variables in the object neighbor are obtained from this vector of identifiers. The order of values in vector `region.id` must be identical to the order of values in any data frames containing information on the spatial lattice, where each row in the data frame corresponds to a single spatial region.

OPTIONAL ARGUMENTS

data if a data frame is associated with the spatial neighbor object, a subset for the data frame may also be obtained. The number of rows in the data frame must be identical to the number of elements in vector `region.id`.

subset a vector of integers or logical values which can be used to index the elements in vector `region.id`.

reorder if TRUE, the variables `row.id`, `col.id`, and `matrix` in the neighbor object are recoded to be sequential integer values.

parameters a vector of parameters which may be present in a spatial model for a covariance structure. If present, these are sorted to correspond to the elements in the component `matrix` of `neighbor`.

weights a vector with length equal to the length of vector `region.id`.

VALUE

a list containing the appropriate subset (or reordering) of the arguments `neighbor`, `data`, `parameters`, and `weights`.

DETAILS

This routine is used to obtain subsets of objects of class `"spatial.neighbor"`, primarily for use with the spatial regression routines.

SEE ALSO

`spatial.neighbor`.

EXAMPLES

```
spatial.subset(sids.neighbor, region.id=sids$id, data=sids, subset=-4)
```

`spatial.sum`   Sum of Two Objects of Class `"spatial.neighbor"`   **`spatial.sum`**

DESCRIPTION

Function used to calculate sum of two objects of class `spatial.neighbor`.

USAGE

```
spatial.sum(neighbor1, neighbor2, parameters1 = NULL,
            parameters2 = NULL)
```

REQUIRED ARGUMENTS

neighbor1 object of class `"spatial.neighbor"` containing the sparse matrix representation of the spatial neighbor matrix (or matrices, see function `spatial.neighbor`).

neighbor2 object of class `"spatial.neighbor"` containing the sparse matrix representation of the spatial neighbor matrix (or matrices, see function `spatial.neighbor`).

OPTIONAL ARGUMENTS

parameters1 a vector with length equal to the number of weight matrices represented by argument `neighbor1`. If `parameters1` is not specified, a vector of ones is assumed.

parameters2 a vector with length equal to the number of weight matrices represented by argument `neighbor2`. If `parameters2` is not specified, a vector of ones is assumed.

VALUE

an object of type `spatial.neighbor` which equals the sum of the two input objects of `spatial neighbor`.

DETAILS

Argument `neighbor` is a sparse matrix representation of one or more matrices using to indicate the strength of "neighbor" relationships for between regions on a spatial grid. Let $a[i,j]$ denote the "(i,j)" element of the k-th such matrix. Then if $a[i,j]$ is not zero, regions i and j are neighbors of the k-th kind, and the strength of the relationship is given by the magnitude of $a[i,j]$. `spatial.weights` computes a weighted sum of the matrices A, where the weights in the linear combination are given by the elements in argument `parameters`.

SEE ALSO

`spatial.neighbor`, `spatial.weights`.

EXAMPLES

```
row.id <- c(1,1,2,2,3)
col.id <- c(2,3,1,3,4)
neighbor1 <- spatial.neighbor(row.id=row.id, col.id=col.id)
col.id <- c(1,2,3,4,1)
neighbor2 <- spatial.neighbor(row.id=row.id, col.id=col.id)
print(spatial.weights(neighbor1))
print(spatial.weights(neighbor2))
a <- spatial.sum(neighbor1,neighbor2)
print(spatial.weights(a))
```

`spatial.weights`	Compute a Spatial Weight Matrix	`spatial.weights`

DESCRIPTION

Given an object of class `"spatial.neighbor"` (and an optional set of neighbor coefficients), `spatial.weights` computes the spatial weight matrix.

USAGE

```
spatial.weights(neighbor, parameters=NULL, region.id=NULL)
```

REQUIRED ARGUMENTS

neighbor an object of class `spatial.neighbor` used (in conjunction with the elements of argument `parameters`) to compute a the spatial weight matrix. In situations with more than one weight matrix (see routine `spatial.neighbor`, and the DETAILS section below), `spatial.weights` computes a weighted sum of the weight matrices.

OPTIONAL ARGUMENTS

parameters a vector with length equal to the number of weight matrices represented by the component `matrix` of argument `neighbor`. If `parameters` is not specified, a vector of all `1`s is assumed.

region.id a vector with length equal to the number of regions in the spatial lattice. If variables `row.id` and `col.id` of argument `neighbor` are not integer valued variables with sequential values from 1 to the number or regions in the lattice, then argument `region.id` must be specified and is used to obtain a sequential coding of the lattice regions.

VALUE

let A_i denote the spatial weights matrix of type i, and let ρ_i denote the corresponding parameter value. Then `spatial.weights` returns $\sum_i \rho_i A_i$.

DETAILS

Argument `neighbor` is a sparse matrix representation of one or more matrices used to indicate the strength of neighbor relationships between regions on a regular or irregular spatial lattice.

Let `a[i,j]` denote a non-zero element of the k-th weight matrix. This implies that regions `i` and `j` are neighbors of the k-th kind, and the strength of this relationship is given by the magnitude of `a[i,j]`.

`spatial.weights` can compute a weighted sum of the matrices `A`, where the weights in the linear combination are given by the elements in argument `parameters`. In practice, you might want to use routine `spatial.weights` when, for example, you want to compute the spatial covariance matrix for a spatial process.

SEE ALSO

`spatial.neighbor`.

EXAMPLES

```
row.id <- c(1,1,2,3)
col.id <- c(2,3,3,4)
neighbor <- spatial.neighbor(row.id=row.id, col.id=col.id, symmetric=T)
spatial.weights(neighbor)
```

spp	Spatial Point Pattern Objects	**spp**

DESCRIPTION

Creates an object of class `"spp"` representing a Spatial Point Pattern. This object is a data frame with columns identifying the locations of the observations. It could contain other columns as well.

USAGE

```
spp(data, x, y, boundary = bbox(data), drop = F)
is.spp(x)
as.spp(x)
```

OPTIONAL ARGUMENTS

data a matrix, or data frame containing the locations that define the spatial point pattern. This could be contained in any two columns of the data frame. If arguments x and y are not used then: if data has columns named `"x"` and `"y"`, these will be understood to contain the corresponding locational information; otherwise, the first two columns will be assumed to contain the necessary information. Use arguments x and y for other options. If data is not specified then x and y must contain the coordinates.

x horizontal coordinate of the spatial point pattern. This argument could be a numeric vector or a character string for different options. If data is given, then x must be a character string or an integer denoting one of the named columns of data. If data is missing, then x must be a vector giving the x-coordinates of the point pattern.

y vertical coordinate of the spatial point pattern. This argument could be a numeric vector or a character string for different options just as x. If data is given, then y must be a character string or an integer denoting one of the named columns of data. This must be different that the one indicated by x. If data is missing, then y must be a vector giving the y-coordinates of the point pattern.

boundary a list with named components `"x"` and `"y"` denoting the vertices of a convex polygon containing the point pattern. The first and last points do not need to coincide. The default is to consider the pattern bounded by a rectangle, its bounding box.

drop logical flag: should the rest of the information (other columns) be left in the resulting data object (which also inherits from class `"data.frame"`) or should the location information be extracted from data? Default is FALSE, that is, leave all other columns intact.

VALUE

the spp function returns an object of class `"spp"`. This is essentially a data frame with two special attributes: attribute `"coords"` which tells corresponding methods which columns denote locational information; and attribute `"boundary"` which contains the same information as entered in the argument boundary.

The is.spp function returns TRUE if object is of class `"spp"` and FALSE otherwise.

The as.spp function coerces a data frame or a matrix with at least two columns to a spatial point pattern object, an object of class `"spp"`.

DETAILS

This function is provided to facilitate calls to methods that act on a spatial point pattern such as the plot method.

The list to be entered as argument boundary can be obtained in several different ways. Use scan to read information that may be available in ASCII files. Use chull or bbox to get the convex hull or bounding box for the point pattern. You may follow those calls by a call to the function poly.expand which will "expand" the polygon by a given fraction keeping its original shape. The list can also be obtained interactively from a graphics device by plotting the x and y coordinates and using the output from locator(type="l") to represent the vertices of the surrounding polygon. If using the latter method, the user must ensure that the resulting polygon is convex, use the function `is.convex.poly` to this end.

SEE ALSO

`plot.spp`, `summary.spp`, `locator`, `chull`, `poly.expand`.

EXAMPLES

```
lansing.m <- spp(lansing[lansing$species=="maple",])

aquifer.xy <- spp(aquifer, drop=T)

mydata <- spp(x = my.x.vector, y = my.y.vector)

#  Given a data frame 'survey' :
#  > names(survey)
#  [1] "a" "b" "y" "x" "c"
#  > attr(spp(survey),"coords")
#  [1] "x" "y"
#  > names(surv.spp <- spp(survey))
#  [1] "a" "b" "y" "x" "c"
```

---

summary.hexbin	Summary Method for a Hexagonally Binned Object	**summary.hexbin**

DESCRIPTION

Returns a list of summary information about an object of class "hexbin".

USAGE

`summary.hexbin(object)`

REQUIRED ARGUMENTS

object data frame object of class `"hexbin"`.

VALUE

table containing the summary of each column in the input object.

SIDE EFFECTS

Prints the call that generated the object, its dimensions and a table with the statistics of each column in the data frame that represents the `"hexbin"` object.

DETAILS

This is the method for objects of class `"hexbin"` of the generic function `summary`.

SEE ALSO

`summary.data.frame`, `hexbin`

EXAMPLES

```
mybin <- hexbin(x,y)
summary(mybin)
```

`summary.slm`	Summary Method for Spatial Linear Models	`summary.slm`

DESCRIPTION

Returns a summary list for a spatial linear model. A null value will be returned if printing is invoked.

USAGE

summary.slm(object, correlation=T)

REQUIRED ARGUMENTS

object the fitted model. This is assumed to be the result of some fit that produces an object inheriting from the class "slm", in the sense that the components returned by the slm function will be available.

OPTIONAL ARGUMENTS

correlation if TRUE, then the variance/covariance and correlation matrices for the coefficients is included in the summary.

VALUE

a list with the following components:

call the function call used to compute the summarized object.

terms the terms in the linear model.

parameters the estimates of the model covariance matrix parameters. A spatial model consists of a specification of the large scale variation (the means), and a specification of the covariance structure of the observed spatial data. Elements in parameters are estimates of parameters used in the specification of this covariance structure.

iterations the total number of iterations used in solving for the covariance matrix parameters.

grad.norm the final gradient norm in the iterations.

objective the minimum objective (minus log likelihood) obtained in the iterations.

message the convergence message.

residuals the residuals from the fit. These are not ordinary residuals. See the cov.family.object help file or the MA, CAR and SAR help files for more information.

coefficients parameters estimated for the linear model, along with standard error estimates, test statistics (that the coefficient is zero), and a two-sided significance level.

sigma the residual standard error estimate.

df the number of degrees of freedom for the model and for residuals.

cov.coef the computed variance/covariance matrix for the estimated linear model coefficients. The parameters estimated for the model covariance matrix are not included.

DETAILS

This function is a method for the generic function summary for class slm . It can be invoked by calling summary for an object of the appropriate class, or directly by calling summary.slm regardless of the class of the object.

SEE ALSO

summary, slm.

EXAMPLES

```
summary(slm(sid.ft~nwbirths.ft, cov.family=SAR, data=sids,
        spatial.arglist=list(neighbor=sids.neighbor)))
```

| `summary.spp` | Summary Method for a Spatial Point Pattern Object | `summary.spp` |

DESCRIPTION

Returns a list of summary information about an object of class `"spp"`.

USAGE

    summary.spp(object, ...)

REQUIRED ARGUMENTS

object data frame object of class `"spp"`.
... other arguments as used by `summary.data.frame`.

SIDE EFFECTS

Prints summary information about `object` such as its total number of points and the extents of its coordinates. If there are other columns in the data frame defining the object, then `summary.data.frame` is called to summarize those.

DETAILS

This function is a method for the generic function `summary` for class spp . It can be invoked by calling `summary` for an object of the appropriate class, or directly by calling `summary.spp` regardless of the class of the object.

SEE ALSO

`spp`, `plot.spp`.

EXAMPLES

    lansing.spp <- spp(lansing)
    summary(lansing.spp)

| `triangulate` | Delaunay's Triangulation | `triangulate` |

DESCRIPTION

Calculate Delaunay's triangulation for points with given coordinates x and y.

USAGE

    triangulate(x, y, plot.it=T, shrink=0.1)

REQUIRED ARGUMENTS

x a list with components `"x"` and `"y"`, a 2-column matrix, or a vector containing the horizontal coordinates of the vertices that form the polygon of interest.

OPTIONAL ARGUMENTS

y if x is a vector of X-coordinates then y must contain the corresponding vertical or Y-coordinates.
plot.it logical flag: should the resulting triangulation be plotted? Default is TRUE.
shrink fraction by which the triangles will be shrunken for better discrimination of the individual triangles in the plot, no edges overlap if shrink > 0.

VALUE

invisibly returns a list with 2 components:

ipt a matrix with 3 rows, for each column the 3 row-values can be used to index x and y and extract corresponding triangle vertices. This provides an ordering of the triangles as well.

ipl another integer matrix with 3 rows. These are the point numbers of the end points of the border line segments and their corresponding triangle number.

SIDE EFFECTS

if `plot.it` = `TRUE` a colorful representation of the triangulation is produced.

DETAILS

A Delaunay triangulation of a point set is a triangulation whose vertices are the point set, with the property that no point in the point set falls in the interior of the circumcircle (circle that passes through all three vertices) of any triangle in the triangulation.

EXAMPLES

```
triangulate(scallops[,c("lat","long")])
```

---

twoway	Fit of a Two-Way Table	**twoway**

DESCRIPTION

Returns a list containing estimated row and column effects as well as a grand effect and the residuals. The default is to give estimates from a median polish. This is a generic function with methods for formula (see `twoway.formula`) and matrices (the default method).

USAGE

```
twoway(x, trim=.5, iter=6, eps=<<see below>>, print=F)
twoway.default(x, trim=.5, iter=6, eps=<<see below>>, print=F)
```

REQUIRED ARGUMENTS

x matrix representing an unreplicated two-way table to be analyzed. Missing values (NAs) are allowed. Rows represent the levels of one of the factors and columns represent the levels of the other factor.

OPTIONAL ARGUMENTS

trim optional trimming fraction for carrying out the analysis. The default value corresponds to using medians. `trim=0` will cause analysis by means, .25 by midmeans, etc.

iter maximum number of full (both row and column) sweeps.

eps error tolerance. If `eps` is given, the algorithm will iterate until the maximum change in row or column effects is less than `eps`. The default is to iterate until the specified number of iterations or until converged to the accuracy of the machine arithmetic. It is not always possible to converge to a unique answer.

print logical flag: if `TRUE`, the maximum change in row/column effects in the last iteration is printed.

VALUE

list with 4 components, `resid`, `row`, `col`, and `grand`, such that
$x[i,j]$ equals `grand` + `row[i]` + `col[j]` + `resid[i,j]`.

grand overall location estimate of the data.

row vector of row effects.

col vector of column effects.

resid matrix of residuals from the fit.

DETAILS

Missing values are omitted in the computations. Results are currently computed to single-precision accuracy only.

With the default `trim=.5` a median polish is performed on the table. Median polish is a simple and very useful technique for detecting anomalous behavior in a two-way table. Although not particularly efficient when the errors have the Gaussian (Normal) distribution, median polish has the highest breakdown possible. Outliers in a two-way (or higher dimensional) table can be very hard to detect without a procedure with at least a moderate breakdown point.

There is no guarantee that median polish will converge for a particular table, but in practice convergence is often achieved after 3 or 4 iterations. Even when there is not convergence, the result will be close to the converged result. Although the median is the L1 (Least Absolute Deviations) solution for the location problem, median polish usually has a sum of absolute deviations greater than an L1 solution, however, it has the good robustness properties that L1 enjoys.

When `trim=0`, a least squares fit is found. If there are no missing values, then this is the fit with which an Analysis of Variance works.

REFERENCES

Hoaglin, D. C., Mosteller, F. and Tukey, J. W., editors (1983). *Understanding Robust and Exploratory Data Analysis.* Wiley, New York.

Mosteller, F. and Tukey, J. W. (1977). *Data Analysis and Regression.* Addison-Wesley, Reading, Mass.

Velleman, P. F. and Hoaglin, D. C. (1981). *Applications, Basics, and Computing of Exploratory Data Analysis.* Duxbury, Boston.

SEE ALSO

`twoway.formula`.
Function `plotfit` produces a graphical display of the fit generated by `twoway`. To simulate a two-way table, see the example in `outer`. `aov` fits an ANOVA model.

EXAMPLES

```
twoway(cereal.attitude, trim=0)   # a two-way ANOVA
twoway(cereal.attitude, trim=.25)  # analysis by midmeans
```

`twoway.formula`	Fit of a Two-Way Table (Formula Method)	`twoway.formula`

DESCRIPTION

Returns a list containing estimated row and column effects as well as a grand effect and the residuals. The default is to give estimates from a median polish.

USAGE

```
twoway.formula(formula=formula(data), data=sys.parent(), subset,
               na.action, trim=.5, iter=6, eps=<<see below>>, print=F)
```

REQUIRED ARGUMENTS

`formula` a formula defining the response and the two predictors. Its form is:

```
z ~ x * y
```

The `z` variable is a numeric response. Missing values (NAs) are allowed. The `x` represents the row variable. The `y` represents the column variable.

OPTIONAL ARGUMENTS

data   an optional data frame in which to find the objects mentioned in `formula`.

subset   expression saying which rows of the data should be used in the fit. This can be a logical vector (which is replicated to have length equal to the number of observations), or a numeric vector indicating which observation numbers are to be included, or a character vector of the row names to be included.

na.action   a function to filter missing data. This is applied to the `model.frame` after any `subset` argument has been used. The default (with `na.fail`) is to create an error if any missing values are found. A possible alternative is `na.omit`, which deletes observations that contain one or more missing values.

trim   optional trimming fraction for carrying out the analysis. The default value corresponds to using medians. `trim=0` will cause analysis by means, .25 by midmeans, etc.

iter   maximum number of full (both row and column) sweeps.

eps   error tolerance. If `eps` is given, the algorithm will iterate until the maximum change in row or column effects is less than `eps`. The default is to iterate until the specified number of iterations or until converged to the accuracy of the machine arithmetic. It is not always possible to converge to a unique answer.

print   a logical flag; if `TRUE`, the maximum change in row/column effects in the last iteration is printed.

VALUE

list with 4 components, `resid`, `row`, `col`, and `grand`, such that
`x[i,j]` equals `grand + row[i] + col[j] + resid[i,j]`

grand   overall location estimate of the data.

row   vector of row effects.

col   vector of column effects.

resid   matrix of residuals from the fit.

call   an image of the call that produced the object.

DETAILS

This is the formula version of `twoway`. The response `z` is converted to a table (matrix) cross classified by the values of `x` and `y`. This matrix is then passed to `twoway.default`. For further details see the `twoway` documentation.

SEE ALSO

twoway.

EXAMPLES

```
twoway(coal ~ x * y, data=coal.ash)
```

---

variogram	Empirical Variogram	**variogram**

DESCRIPTION

Computes the empirical variogram for two dimensional spatial data. Multiple variograms for different directions can be computed.

USAGE

```
variogram(formula, data=<<see below>>, subset=<<see below>>,
        na.action=<<see below>>, lag=<<see below>>, nlag=20,
        tol.lag=lag/2, azimuth=0, tol.azimuth=90, bandwidth=1e21,
        maxdist=<<see below>>, minpairs=6, method="classical")
```

REQUIRED ARGUMENTS

formula   formula defining the response and the predictors. In general, its form is:

```
z ~ x + y
```

The z variable is a numeric response. Variables x and y are the locations. All variables in the formula must be vectors of equal length with no missing values (NAs). The formula may also contain expressions for the variables, e.g. `sqrt(count)`, `log(age+1)` or `I(2*x)`. (The `I()` is required since the `*` operator has a special meaning on the right side of a formula. The right hand side may also be a call to the `loc` function e.g. `loc(x,y)`. The `loc` function can be used to correct for geometric anisotropy, see the `loc` help file.

OPTIONAL ARGUMENTS

    data   an optional data frame in which to find the objects mentioned in `formula`.

  subset expression saying which subset of the rows of the data should be used in the fit. This can be a logical vector (which is replicated to have length equal to the number of observations), or a numeric vector indicating which observation numbers are to be included, or a character vector of the row names to be included.

na.action a function to filter missing data. This is applied to the `model.frame` after any `subset` argument has been used. The default (with `na.fail`) is to create an error if any missing values are found. A possible alternative is `na.omit`, which deletes observations that contain one or more missing values.

    lag a numeric value, the width of the lags. If missing, `lag` is set to `maxdist / nlag`.

   nlag an integer, the maximum number of lags to calculate.

 tol.lag a numeric value, the distance tolerance.

 azimuth a vector of direction angles in degrees, measured clockwise from North. A separate variogram will be estimated for each direction.

tol.azimuth angle tolerance in degrees. A `tol.azimuth` of 90 or greater (the default) results in an omnidirectional variogram.

bandwidth the maximum bandwidth, the deviation from the direction orthogonal to the direction angle.

 maxdist the maximum distance to include in the returned output. The default is half the maximum distance in the data.

minpairs the minimum number of pairs of points (minimum value for np) that must be used in calculating a variogram value. If np is less than `minpairs` then that value is dropped from the variogram.

 method a character string to select the method for estimating the variogram. The possible values are `"classical"` for Matheron's (1963) estimate and `"robust"` for Cressie and Hawkins (1980) robust estimator. Only the first character of the string needs to be given.

VALUE

an object of class `"variogram"` that inherits from `"data.frame"` with columns:

 distance the average distance for pairs in the lag.

  gamma the variogram estimate.

    np the number of pairs in each lag.

 azimuth a factor denoting the angular direction.

The return object has an attribute `call` with an image of the call that produced the object.

DETAILS

Method for class `"variogram"` include `plot`, `print` and `summary`.

The variogram is a measure of spatial correlation. This function uses a modified version of the GSLIB subroutine gamv2 (Deutsch and Journel, 1992).

REFERENCES

Cressie, Noel A. C. (1993). *Statistics for Spatial Data,* Revised Edition. Wiley, New York.

Cressie, N. and Hawkins, D. M. (1980). Robust estimation of the variogram. *Mathematical Geology* **12**, 115-125.

Deutsch, Clayton V. and Journel, Andre G. (1992). *GSLIB Geostatistical Software Library and User's Guide.* Oxford University Press, New York.

Matheron, G. (1963). Principles of geostatistics. *Economic Geology* **58**, 1246-1266.

SEE ALSO

correlogram, loc, plot.variogram.

EXAMPLES

```
# an omnidirectional variogram
v1 <- variogram(ore ~ easting + northing, data=iron.ore)
plot(v1)
# variograms in 0, 45, 90 and 135 degrees directions
v2 <- variogram(ore ~ loc(easting,northing), data=iron.ore,
          azimuth=c(0,45,90,135), tol.azimuth=22.5)
plot(v2)
```

---

variogram.cloud	Calculate Variogram Cloud	**variogram.cloud**

DESCRIPTION

Calculates all pairwise differences in a random field data set.

USAGE

```
variogram.cloud(formula, data=<<see below>>, subset=<<see below>>,
          na.action=<<see below>>, azimuth=0, tol.azimuth=90,
          maxdist=<<see below>>, bandwidth=1e+307,
          FUN=function(zi, zj) (zi - zj)^2/2))
```

REQUIRED ARGUMENTS

formula   formula defining the response and the predictors. In general, its form is:

z ~ x + y

The z variable is a numeric response. Variables x and y are the locations. All variables in the formula must be vectors of equal length with no missing values (NAs). The formula may also contain expressions for the variables, e.g. sqrt(count) or log(age+1). The right hand side may also be a call to the loc function e.g. loc(x,y). The loc function can be used to correct for geometric anisotropy, see the loc help file.

OPTIONAL ARGUMENTS

data   an optional data frame in which to find the objects mentioned in formula.

subset   expression saying which subset of the rows of the data should be used in the fit. This can be a logical vector (which is replicated to have length equal to the number of observations), or a numeric vector indicating which observation numbers are to be included, or a character vector of the row names to be included.

na.action   a function to filter missing data. This is applied to the model.frame after any subset argument has been used. The default (with na.fail) is to create an error if any missing values are found. A possible alternative is na.omit, which deletes observations that contain one or more missing values.

azimuth   the clockwise direction angle in degrees from North-South. Only pairs of points in this direction plus or minus tol.azimuth will be included in the output.

tol.azimuth   the tolerance angle, in degrees. tol.azimuth greater than or equal to 90 implies the of use all directions.

maxdist   the maximum distance to consider. The default is half the maximum observed distance.

bandwidth   the maximum perpendicular distance to consider.

FUN   a function of two variables that is to be computed. The default function is the contribution to the classical empirical variogram for the pair `z[i]`, `z[j]`.

VALUE

an object of class `"vgram.cloud"` that inherits from `"data.frame"`. The columns are:

distance   the distance between the two points.

gamma   the value of FUN for the `z[iindex]`, `z[jindex]`.

iindex   the index into the original data for the first value of the pair.

jindex   the index into the original data for the second value of the pair.

The return object has an attribute `call` with an image of the call that produced the object.

DETAILS

Methods for class `"vgram.cloud"` include `boxplot`, `plot` and `identify`.

If all directions and distances are included the return object will have n*(n-1)/2 rows where n is the number of observations. This can get very large, even for relatively small n. The argument `maxdist` can be used to limit the size. Typically values beyond half the maximum distance in the data are not used in estimating the variogram function.

REFERENCES

Cressie, Noel. (1993). *Statistics For Spatial Data,* Revised Edition. Wiley, New York.

SEE ALSO

`boxplot.vgram.cloud`, `identify.vgram.cloud`, `plot.vgram.cloud`, `variogram`.

EXAMPLES

```
v1 <- variogram.cloud(coal ~ x + y, data=coal.ash)
plot(v1)
boxplot(v1)
```

---

write.geoeas	Write A GEO-EAS Data File	**write.geoeas**

DESCRIPTION

Writes out a data frame as a GEO-EAS data file.

USAGE

```
write.geoeas(data, file="", title, units, sep="\t",
             end.of.row="\r\n")
```

REQUIRED ARGUMENTS

data   a data frame.

OPTIONAL ARGUMENTS

file   the name of the date file to be created. If `file` is "" then `write.geoeas` will write to standard output.

title   the descriptive title for the GEO-EAS data set. Can be up to 80 characters. If not supplied the name of the `data` argument is used.

units   optional units vector. This must be the same length as the number of columns in `data`.

sep   the separator character to use between data fields in the file. The default is the tab character.

end.of.row   the character to print at the end of each row of the file. The default is carriage-return-newline since GEO-EAS data files are typically used on PC's.

SIDE EFFECTS

A GEO-EAS data file is written.

DETAILS

GEO-EAS is a collection of MS-DOS based software tools for performing two-dimensional geostatistical analysis of spatially distributed software. It uses a fixed file format with header information for its data input files. Many other PC based spatial analysis programs also use this file format. See Englund and Sparks (1992) for more details.

REFERENCES

Englund, Evan and Sparks, Allen. (1992). GEO-EAS: Geostatistical Environmental Assessment Software Users's Guide, Version 1.2.1. GWMC-FOS 53 PC, International Ground Water Modeling Center, Golden, CO.

SEE ALSO

```
read.geoeas, write.table.
```

EXAMPLES

```
# Create a GEO-EAS file from the iron.ore data frame
write.geoeas(iron.ore, "iron.dat")
```

---

xy2cell	Compute Hexagon Cell Id's From x and y	**xy2cell**

DESCRIPTION

Computes hexagon cell id's corresponding to xy-pairs. Used for partitioning data into hexagonal regions and for computation of statistics for each of these regions. Provides symbol congestion control in maps.

USAGE

```
xy2cell(x, y, xbins=30, shape=1, xlim=range(x), ylim=range(y))
```

REQUIRED ARGUMENTS

x   numeric vector. Usually the first (horizontal) coordinate of bivariate data to be binned into hexagonal cells.

y   numeric vector. Usually the second (vertical) coordinate of bivariate data to be binned.

OPTIONAL ARGUMENTS

xbins   number of hexagonal cells partitioning the range of x values.

shape   height to width ratio for the hexagonal cells.

xlim   the horizontal limits of the binning region in units of x. By default these are the minimum and maximum values of x.

ylim   the vertical limits of the binning region in y units. This defaults to the minimum and maximum values of y.

VALUE

a vector of cell identifiers that can be mapped into the bin centers in data units. This vector will have the same length as x and y. The result also has the following attributes:

xbins   number of hexagonal cells across the x axis.

shape   same as the input parameter shape.

xlim   same as input parameter xlim.

ylim   same as input parameter ylim.

DETAILS

The plot shape must be maintained for hexagons to appear with equal sides. Calculations are in single precision.

This function can be used to compute statistics per cell. See the EXAMPLES below for one such instance.

REFERENCES

Carr, D. B., Olsen, A. R. and White, D. (1992). Hexagon mosaic maps for display of univariate and bivariate geographical data. *Cartography and Geographics Information Systems,* **19**, 228-236.

SEE ALSO

```
cell2xy, hexbin, summary.hexbin, plot.hexbin, identify.hexbin, smooth.hexbin,
erode.hexbin.
```

EXAMPLES

```
ozone.bin <- hexbin(ozone.xy$x, ozone.xy$y, xbins=8)
ozone.cells <- xy2cell(ozone.xy$x, ozone.xy$y, xbins=8)
# Find the median of each hexagonal cell:
ozone.angle <- tapply(ozone.median,ozone.cells,median)
map(region=c("new york","new jersey","conn","mass"),lty=2)
rayplot(ozone.bin$xcenter,ozone.bin$ycenter,ozone.angle)
```

Glossary

Anisotropic process Spatial process in which similarities between data at paired locations dependent on distance and direction between the two locations.

Areal referenced data Data indexed by region rather than by specific location.

Azimuth Angular distance measured on a planar circle in a clockwise direction from the north.

Correlogram A ratio of covariances with increasing distance.

Convex hull The minimal polygon enclosing all points, which can be found by triangulating the set of points and taking the outermost edges of the triangulation.

Geostatistical data Spatial data where observations are recorded at fixed locations throughout a continuous region.

Hexagonal binning A procedure by which locational data are grouped into cells according to the dimensions of a set of hexagonal tessellations.

Intensity The number of points per unit area for spatial point patterns.

Intrinsic stationarity Property of a spatial process that requires the variance of *increments* to be independent of location, where an increment is defined as the first-order difference between two points. *See Stationarity*.

Isotropic process Spatial process in which similarities between data at paired locations dependent only on distance, and not on the direction between the two locations.

K-function An alternative to second-order intensity to evaluate the second-order properties (spatial dependence) of a spatial point pattern.

Kriging A linear interpolation method for predicting geostatistical data values at locations which have not been observed. Incorporates a variogram model for spatial correlation.

Lattice data Spatial data with observations associated with irregularly or regularly spaced spatial regions.

Local stationarity Refers to some form of *location invariance* of the data in subregions, where location invariance implies that the relationships between any subset of points remain the same no matter where the points reside in the subregion.

Loess Local regression model.

Nearest-neighbor distances The set of distances between each point in a spatial point pattern and its next closest point.

Neighbor matrix A square matrix with rows and columns representing regions. For each row, columns containing positive values are neighbors of the region represented by that row. The relative size of the cell value (the *neighbor weight*) indicates the strength of the neighbor relationship.

Neighbor weights The values in the cells of the neighbor matrix. These values are used to weight the correlation between neighbors. They are separate and distinct from the weight matrix used to correct for nonhomogeneous variance.

Point patterns *See* Spatial point patterns.

Point referenced data Data indexed by exact data locations.

Random field data *See Geostatistical data*

Ripley's K-function An estimator of the theoretical K-function for spatial point pattern data, which accounts for edge effects.

Second-order intensity A measure of spatial dependence of a spatial point pattern used to detect clustering or regularity.

Semivariogram One-half of the variogram; *see Variogram.*

Spatial autocorrelation Lack of independence in spatial data. Occurs when the data values in neighboring sites or regions vary together.

Spatial data A collection of measurements or observations taken at specific locations or within specific regions.

Spatial dependence Data correlation or covariance between nearby or neighboring sites or regions.

Spatial neighbors Two sites with spatially correlated data values. In practice, these are often defined as adjacent sites or sites within a certain distance of each other.

Spatial point patterns Spatial data with spatial locations as the variable to be analyzed.

Spatial regression A method of modeling lattice data using a linear model for large scale trends and a covariance model for small-scale variation.

Spatial trend A large-scale change in the mean through spatial location.

Stationarity Refers to some form of *location invariance* of the data, where location invariance implies that the relationships between any subset of points remain the same no matter where the points reside in space.

Strict stationarity Requires an equivalence of distribution functions under translation and rotation. *See Stationarity.*

Symmetric neighbors A pair of neighbors with the property that the presence and strength of the relationship between the two is equivalent in both directions. In the context of a neighbor matrix, the neighbor relationships are symmetric if the value in cell (i, j) is identical to the value in cell (j, i).

Tesselation An aggregate of cells that covers space without overlapping.

Variogram A measure of the similarity between pairs of points with increasing distance. The variogram is based on the average squared differences between all pairs.

Variogram cloud A plot of the squared differences between points versus the separation distance.

Weak stationarity Requires a constant mean and a covariance that is independent of location. The covariance is only dependent on the distance (and perhaps direction) between points. Requires the existence of a positive, finite variance. *See Stationarity.*

Weight matrix A diagonal matrix sometimes used to correct for nonhomogeneous variance in a spatial regression model. For example, if the variance is dependent on sample size, the values on the diagonal might be inverted sample sizes. The weight matrix is separate and distinct from the neighbor matrix, which contains neighbor weights.

Bibliography

Bates, D. M. and Chambers, J. M. (1992). Nonlinear models. In Chambers, J. M. and Hastie, T. J., editors, *Statistical Models in S*, pages 421–453. Wadsworth and Brooks, Pacific Grove, California.

Carr, D. B., Olsen, A. T., and White, D. (1992). Hexagon mosaic maps for display of univariate and bivariate geographical data. *Cartography and Geographical Information Systems*, 19:228–236.

Cleveland, W. S., Grosse, E., and Shyu, W. M. (1992). Local regression models. In Chambers, J. M. and Hastie, T. J., editors, *Statistical Models in S*, pages 309–376. Wadsworth and Brooks, Pacific Grove, California.

Cliff, A. D. and Ord, J. K. (1981). *Spatial Processes: Models and Applications*. Pion Limited, London.

Cressie, N. (1985). Fitting variogram models by weighted least squares. *Mathematical Geology*, 17:563–586.

Cressie, N. (1986). Kriging nonstationary data. *Journal American Statistical Association*, 81:625–634.

Cressie, N. (1989). Geostatistics. *American Statistician*, 43:197–202.

Cressie, N. and Chan, N. H. (1989). Spatial modeling of regional variables. *Journal American Statistical Association*, 84:393–401.

Cressie, N. and Hawkins, D. M. (1980). Robust estimation of the variogram: I. *Mathematical Geology*, 12:115–125.

Cressie, N. and Read, T. R. C. (1985). Do sudden infant deaths come in clusters. *Statistics and Decisions, Supplemental Issues No. 2*, 3:333–349.

Cressie, N. and Read, T. R. C. (1989). Spatial data analysis of regional counts. *Biometrical Journal*, 31:699–719.

Cressie, N. A. C. (1993). *Statistics for Spatial Data*. Wiley, New York.

Diggle, P. J. (1983). *Statistical Analysis of Spatial Point Patterns*. Academic Press, New York.

Ecker, M. D. and Heltshe, J. F. (1994). Geostatistical estimates of scallop abundance. In Lange, N., Ryan, L., Billard, L., Brillinger, D., Conquest, L., and Greenhouse, J., editors, *Case Studies in Biometry*, pages 107–124. Wiley, New York.

Englund, E. and Sparks, A. (1992). GEO-EAS: Geostatistical Environmental Assessment Software. User's Manual IGWMC-FOS 53 PC, International Ground Water Modeling Center, Golden, CO. Version 1.2.1.

Gatrell, A. C., Bailey, T. C., Diggle, P. J., and Rowlingson, B. S. (1995). Spatial point pattern analysis and its application in geographical epidemiology. Research Report 29, North West Regional Research Laboratory, Lancaster University, Lancaster, UK.

Gerrard, D. J. (1969). Competition quotient: a new measure of the competition affecting individual forest trees. Research Bulletin No. 20, Agricultural Experiment Station, Michigan State University.

Gomez, M. and Hazen, K. (1970). Evaluating sulfur and ash distribution in coal seams by statistical response surface regression analysis. Technical Report RI7377, U.S. Bureau of Mines.

Griffith, D. A. (1995). Some guidelines for specifying the geographic weights matrix contained in spatial statistical models. In Arlinghaus, S. L., editor, *Practical Handbook of Spatial Statistics*, pages 65–82. CRC Press, Inc., Boca Raton.

Haining, R. (1990). *Spatial Data Analysis in the Social and Environmental Sciences*. Cambridge University Press, Cambridge.

Harper, W. V. and Furr, J. M. (1986). Geostatistical analysis of potentiometric data in the Wolfcamp Aquifer of the Palo Duro Basin, Texas. Technical Report ONWI-587, Batelle Memorial Institute, Columbus, Ohio.

Hutchings, M. J. (1979). Standing crop and pattern in pure stands of *Mercurialis perennis* and *Rubus fruticosus* in mixed deciduous woodland. *Oikos*, 31:351–357.

Isaaks, E. H. and Srivastava, R. M. (1989). *An Introduction to Applied Geostatistics*. Oxford University Press, New York.

Journel, A. G. and Huijbregts, C. J. (1978). *Mining Geostatistics*. Academic Press, London.

Matheron, G. (1963). Principles of geostatistics. *Economic Geology*, 58:1246–1266.

Ripley, B. D. (1976). The second-order analysis of stationary point processes. *Journal of Applied Probability*, 13:255–266.

Ripley, B. D. (1981). *Spatial Statistics*. Wiley, New York.

Symons, M. J., Grimson, R. C., and Yuan, Y. C. (1983). Clustering of rare events. *Biometrics*, 39:193–205.

Venables, W. N. and Ripley, B. D. (1994). *Modern Applied Statistics with S-PLUS*. Springer-Verlag, New York.

Zimmerman, D. L. and Zimmerman, M. B. (1991). A comparison of spatial semivariogram estimators and corresponding ordinary kriging predictors. *Technometrics*, 33:77–92.

Index

INDISPENSABLE S-PLUS RESOURCES!

Second Edition
W.N. VENABLES and **B.D. RIPLEY**
MODERN APPLIED STATISTICS WITH S-PLUS

This book is a guide to using S-PLUS to perform statistical analyses and provides both an introduction to the use of S-PLUS and a course in modern statistical methods. It shows how to use S-PLUS as a powerful and graphical system, providing an indepth coverage of its use on the Windows and UNIX platforms. Throughout, the emphasis is on presenting practical problems and full analyses of real data sets. Many of the methods discussed are state-of-the-art approaches to topics such as linear and non-linear regression models, robust and smooth regression methods, survival analysis, multivariate analysis, tree-based methods, time series, spatial statistics, and classification. This edition is intended for users of S-PLUS 3.3, 3.4, 4.0, or later, and covers the recent developments in graphics, the new statistical functionality, and the new object-oriented programming features. The authors have written several software libraries which enhance S-PLUS; these and all the datasets used are available on the Internet. There are also on-line complements covering advanced material, further exercises, and new features of S-PLUS as they are introduced.

1997/APP. 568 PP./HARDCOVER/ISBN 0-387-98214-0

ANDREAS KRAUSE and **MELVIN OLSON**
THE BASICS OF S AND S-PLUS

This book explains the basics of S-PLUS in a clear style–at a level suitable for people with little computing or statistical knowledge. Unlike the S-PLUS manuals, it is not comprehensive, but instead introduces the most important ideas of S-PLUS through the use of many examples. Each chapter includes a collection of exercises, accompanied by fully worked-out solutions and detailed comments. The volume is rounded off with practical hints on how efficient work can be performed in S-PLUS.

1997/APP. 272 PP., 34 ILLUS./SOFTCOVER
ISBN 0-387-94985-2

ANDREW BRUCE and **HONG-YE GAO**
APPLIED WAVELET ANALYSIS WITH S-PLUS

This book introduces applied wavelet analysis through the S-PLUS software system. Using a visual data analysis approach, wavelet concepts are clearly explained. In addition to wavelets, a whole range of related signal processing techniques such as wavelet packets, local cosine analysis, and matching pursuits are covered. Applications of wavelet analysis are illustrated, including nonparametric function estimation, digital image compression, and time-frequency signal analysis. The book is intended for a broad range of data analysts, scientists, and engineers.

1996/338 PP./SOFTCOVER/ISBN 0-387-94714-0

TO ORDER OR FOR INFORMATION:

In North America: Call 1-800-SPRINGER
Springer-Verlag New York, Inc., 175 Fifth Avenue, New York, NY 10010
WWW: http://www.springer-ny.com • Email: orders@springer-ny.com

Outside North America: Call: +49/30/8 27 87-3 73 • +49/30/8 27 87-0
Springer-Verlag, P.O. Box 31 13 40, D-10643 Berlin, Germany
WWW: http://www.springer.de • Email: orders@springer.de

Springer

#S293A